674.4/RUD

Mach

Machine Woodworking

Nick Rudkin

Huddersfield Technical College

A member of the Hodder Headline Group
LONDON • SYDNEY • AUCKLAND

First published in Great Britain 1998 by Arnold
a member of the Hodder Headline Group,
338 Euston Road, London NW1 3BH
http://www.arnoldpublishers.com

British Library Cataloguing in Publication Data
A catalogue record for this book is available from the British Library

ISBN: 0 340 61423 4 ♪

1 2 3 4 5 6 7 8 9 10

Publisher: Eliane Wigzell
Production Editor: Liz Gooster
Production Controller: Priya Gohil
Cover Design: Doodles

Typeset in 10.5/11.5 Plantin by J&L Composition Ltd, Filey, North Yorkshire
Printed and bound in Great Britain by J.W. Arrowsmith Ltd., Bristol

Contents

Acknowledgements

Grateful appreciation is expressed to the following individuals and organizations for help, assistance or advice during the writing of this book:

Lietz Tooling UK Ltd
Wadkin Ltd
Stenner of Tiverton
Whitehill Tooling Ltd
Construction Industry Training Board
Health and Safety Executive
Robin Davies

Introduction: The Law

For many years the woodworking machines regulations of 1974 have been the machine woodworker's 'bible', giving the legal requirements for setting guards and operating the equipment. In January of 1991 these regulations were revoked to make way for a new era of safety awareness. Known as PUWER (provision and use of work equipment regulations), a booklet was issued which details all the areas of safety relating to work equipment. It is for all industries, not just wood machining, and therefore is not specific to actual machines. For this reason the Health and Safety Executive have produced a series of woodworking information sheets that show and explain the most suitable and, more importantly, the safest ways to set up and operate machines.

The new regulations set out a four-step procedure where the operator must try to achieve the safety standards specified in each step. The four steps are:

1. Use fixed enclosing guards. Basically this means any standard guarding arrangement on a machine must be correctly set. This would be a bridge guard on a surface planer when used for planing a face and edge on to a piece of timber.
2. If the standard guard cannot be fitted or so adjusted to prevent entry to the danger area, then an extra guard or safety device must be fitted. The bridge guard mentioned in (1) above would be of little use when cutting a rebate on a surface planer. In this case a tunnel guard would be used.
3. To prevent entry to the danger area 'protection appliances' may be needed. The rebating of timber on a surface planer will require the use of a push stick to assist in feeding the material out of the machine. Jigs, bed pieces, saddles and templates all come under the heading of protection appliances.
4. Finally information, instruction, training and supervision must be provided for all operatives on all operations. This could be in-house training where the company appoints personnel to carry out training and instruction to machine operators or college training leading to a nationally (or internationally) recognised qualification.

To the trained or experience machinist these regulations are simple to adhere to, relying on their skill and previous training to set guards safely. To a new machinist however these new regulations give no indication of what is actually safe. One person may consider a guard to be safe if it is 20 mm away from a component. The next operator may think 50 mm for the same job. For this reason a whole regulation (Regulation 9, *Training*) has been dedicated to the requirement of training. This also makes reference to the fact that the requirement for training in the 1974 regulations must continue to apply.

Once machinists are considered to be qualified and therefore competent to the point of being able to set up and work a machine for a specific task, then they become responsible for their own safety when working and also for that of others around them who may be affected by their work. Any breaches of regulations or unsafe working practices are ultimately their own doing and responsibility lies with them. Inexperienced operatives should first of all be well instructed and trained to carry out the job in question. Second, adequate supervision must be given to prevent any hazards before work commences.

This book is designed to give a good grounding to inexperienced machine operators in all trade areas (machine woodwork, carpentry and joinery and furniture industries) by explaining the theoretical and practical aspects of machining. It will also provide a valuable reference to the more experienced operative. The techniques used in this book are deemed to be safe, acceptable and in compliance with national regulations but it must be realized that there is always more than one acceptable way of doing something.

1 Circular sawing machines

1.1 Circular rip saw construction

1.1.1 Basic functions

To convert or rip timber from its marketable size into sections required to comply with a cutting list.

1.1.2 Design and layout

A wide variety of machines, made by different manufacturers, are available which may vary in design and layout. Safety devices in compliance with current regulations may well be standard, but adjustments to these devices and other controls may be different. It is therefore advised that a study of the machine is made prior to operating (**Fig. 1.1**).

1.1.3 Types of saw blade

Parallel plate saw blades are the most common type of blade used on a circular rip saw in the construction industry.

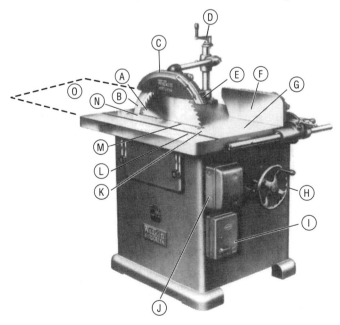

A - Saw blade
B - Riving knife
C - Crown guard
D - Crown guard adjustment
E - Extension guard
F - Adjustable fence (will tilt up to 45°)
G - Table
H - Handwheel – saw spindle rise and fall
I - Isolator
J - Control – start and stop buttons
K - Mouthpiece (hardwood)
L - Machine groove for cross-cutting gauge
M- Saw blade packing
N - Finger plate (access to saw spindle)
O - Extension table (provision to comply with Regulation 20(2))

Fig. 1.1 Circular rip sawing machine.

They can be spring set, where the teeth are bent in alternate directions to give clearance to the body of the blade, or have tungsten carbide tips (TCT) welded on to the outside edge.

1.1.4 Spring set blades

Set is an important factor to be considered for efficient running of the machine. Blades require as little set as possible, while still giving clearance (**Fig. 1.2**).

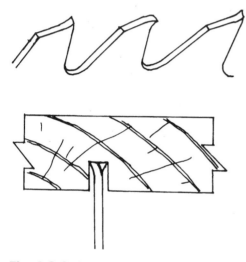

Fig. 1.2 Spring set blade.

Too much set will result in increased waste, increased power consumption, excessive operator fatigue and often a poor finish. Too little set must also be avoided because, once cut, the timber may bind on the blade body and cause thermal stress (overheating) in the blade. The timber may also be violently thrown back as it makes contact with the up-running teeth.

One of the disadvantages of spring set plate saw blades is that when cutting abrasive timbers or manufactured sheet materials (chipboard, plywood, medium density fibreboard (MDF)) the sharp edge of the tooth point is quickly lost, making the blade blunt. The use of tungsten carbide tipped blades is recommended for this type of work.

1.1.5 Tungsten carbide tipped blades

Tungsten carbide is an extremely hard material used as a hard facing for machine tools. The saw plate is pressed out to leave seatings for small tips to be heat fused into place (**Fig. 1.3**). The teeth are shaped in moulds during their manufacture to give clearance or set to the blade. TCT blades are usually more efficient because all the teeth cut to both sides of the material. On a blade with 60 teeth, all teeth will do work on both sides of the cut. A spring set blade with 60 teeth will have 30 teeth working to the left and 30 to the right.

Due to the hardwearing nature of tungsten carbide the tipped blades far outlast spring set blades. Even dense or abrasive hardwoods and sheet materials have little effect on their cutting edge.

Used on a far smaller scale nowadays, and usually confined to the furniture industry are the types shown in **Figs 1.4–1.6**. They are all used for cutting expensive sheet materials and hardwoods where their fine saw cut gives less waste. They are, however, restricted to small depths of cut.

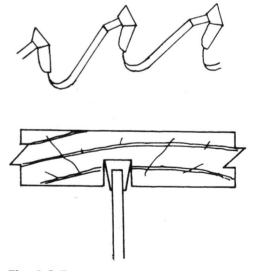

Fig. 1.3 Tungsten carbide tipped blade.

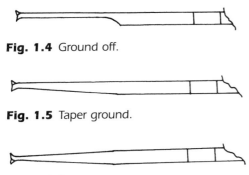

Fig. 1.4 Ground off.

Fig. 1.5 Taper ground.

Fig. 1.6 Swage.

1.1.6 Teeth

The teeth on a circular saw blade determine the cutting efficiency, and vary in shape depending on the material being cut. Three angles form the tooth shape and, by altering these angles, the cutting action is changed (**Fig. 1.7**).

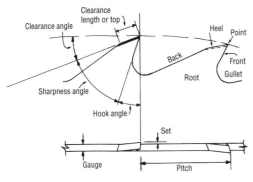

Fig. 1.7 Angles and parts of a softwood tooth.

1.1.7 Hook angle

Softwoods such as Redwood (Scots pine) will require a hook angle of 30°. This allows a sharp point to chisel into the wood and at the same time a large gullet area to collect waste without blocking up. Although the point is sharp it is also weak. Softwoods present no problem to this but, because some hardwoods are so dense and abrasive, they could cause the teeth to break off or, more commonly, produce a severe blunting effect. To overcome this problem the hook angle is made smaller (between 5° and 20° depending on the species) which allows the point still to lead the cut but gives more strength to the teeth. Sacrificed for the strength is the gullet area which, due to being smaller, will mean slower feed speeds.

1.1.8 Sharpness angle

The sharpness angle is directly related to the hook angle. It is measured from the top of a tooth to the front, and as the hook angle is made bigger the sharpness angle becomes smaller. The smaller the sharpness angle, the sharper the tooth point. A softwood tooth with a big hook angle will be much sharper but not as strong as a hardwood tooth (**Fig. 1.8**).

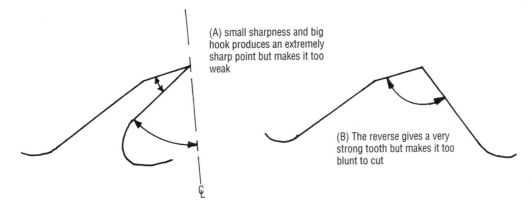

(A) small sharpness and big hook produces an extremely sharp point but makes it too weak

(B) The reverse gives a very strong tooth but makes it too blunt to cut

Fig. 1.8 Effects of exaggerated hook and sharpness angles.

1.1.9 Finger plate, packings and mouthpiece

The finger plate is fitted around the blade. It can be removed to give access to the blade for maintenance and servicing. Gaps are left in the finger plate next to the blade to allow packings to be fitted. These packings are usually made of felt, pressed into the finger plate to give support to the blade as it runs. By lubricating the packings with a mixture of paraffin and oil, the blade will not build up with resin from the timber and cooler cutting will result. It must be noted that the packings must not extend any further forward than the gullets of the teeth or the set could chew them up. To help keep the packings in place, a mouthpiece is fitted around the teeth. The mouthpiece should be a good fit to prevent splinters of wood getting between the blade and table. Several different mouthpieces may be required to suit different blade diameters (**Fig. 1.9**).

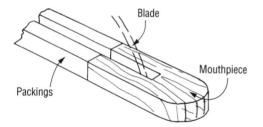

Fig. 1.9 Mouthpiece and packings.

1.1.10 Riving knife

The riving knife is a curved piece of steel directly behind the blade. It serves two important purposes:

- it prevents timber from binding on the blade;
- and it acts as a guard to the back of the blade.

It should be thicker than the body of the blade but thinner than the kerf – the term given to the width of cut produced by the blade (body plus set; **Fig. 1.10**).

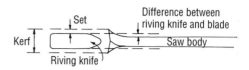

Fig. 1.10 *Plan view of blade and riving knife.*

For setting and positioning the riving knife, adhere to **Figs 1.11–1.13** with reference to current regulations.

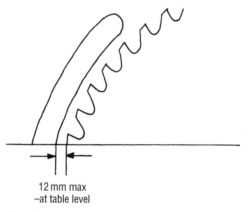

12 mm max
–at table level

Fig. 1.11 Distance from blade.

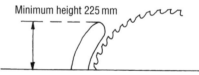

Minimum height 225 mm

Fig. 1.12 Height of riving knife for blades over 600 mm.

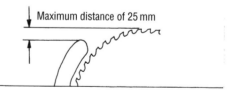

Maximum distance of 25 mm

Fig. 1.13 Height of riving knife for blades of 600 mm or less.

1.1.11 Fence positioning

It is important for the fence to be set to suit the diameter and height of the blade in use. This is to prevent the timber from binding on the blade and to achieve the best sawn finish possible (**Fig. 1.14**).

In the figure, position A is too far for-

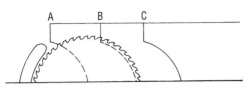

Fig. 1.14 Positioning the fence.

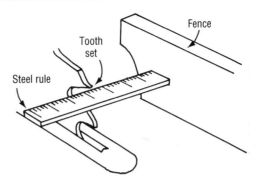

Fig. 1.16 Setting width of cut.

ward, the timber will bind on the blade after completion of cut.

Position B is correct – the fence gives guidance through the cut and the timber is clear on completion.

In position C the timber will leave the fence prior to completion. This will allow the timber to wander, leaving an erratic edge that could present problems when planing.

In order to prevent the up-cutting teeth from scoring the cut surface, a very slight lead (or runout) is applied to the fence. This will mean that, as the timber travels to the far side of the blade, it will have a clearance from the teeth. This will also prevent binding (**Fig. 1.15**).

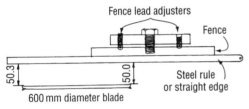

Fig. 1.15 Lead applied to fence.

1.2 Using a circular rip saw

1.2.1 Setting up

Most machines will be fitted with a scale for setting the width of cut. On older machines where the scale is damaged or not properly calibrated, the use of a tape measure or steel rule will make setting easier (**Fig. 1.16**).

1. Isolate the machine.
2. Position rule against the fence with graduations extending towards the

blade and adjust the position of the fence to give desired width of cut. *Note* Use a tooth that has been set towards the fence.
3. Adjust the blade height to suit the type of work being done. For example, if cutting timber on a saw bench prior to planing, the blade can be set high to give a more aggressive cut. This will allow faster feed speeds and less wear on the cutting edges because the blade is not inside the timber for any great length of time. It will, however, produce a rough finish on the underside of the timber (breakout) as the teeth smash through.

Figure 1.17(a) shows the blade positioned with the roots of the blade just clearing the material. It can be seen that at any one time, more blade is contained in the material than in **Fig. 1.17(b)**.

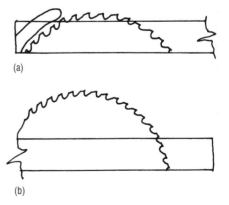

Fig. 1.17 Blade heights.

This will produce more of a shearing cut, thus producing a better finish on the cut surface as well as reducing breakout on the underside. This is particularly useful when cutting sheet materials or when a sawn finish will be seen on the finished job.

4. Ensure that the fence is in line with the gullets of the blade.
5. Lower the crown guard into position making sure that:

   (a) the gullets of the teeth projecting through the material are concealed within the guard; and
   (b) the guard is not more than 12 mm above the material. On some machines an extension guard is fitted to the crown guard. If this is the case and it is 12 mm from the material then the crown guard can be set higher, provided that it conceals the gullets (**Fig. 1.18**).

6. Check all parts are locked and proceed to run a test cut.

1.2.2 Cutting material

1. The first point to consider, before starting the machine, is 'where is the most practical place to stand when feeding?' The answer must allow efficient feeding and working but, most importantly, it must be safe (**Fig. 1.19**).
2. It is illegal to feed a circular sawing machine without the aid of a push stick. As a general rule a push stick should be between 500 and 600 mm long with a comfortable handle so as to encourage its use (**Fig. 1.20**).

   It is a requirement that a push stick is used to exert feeding pressure for the last 300 mm of cut or the entire cut if the material is less than 300 mm. It must also be used to remove offcuts and waste from the machine.

   A push stick can only be dispensed with when:

   (a) the distance between the blade and fence is so big that feeding by hand can be carried out safely;
   (b) when using an automatic feeding unit or carriage; and
   (c) when using a sliding table.

3. The feeding of cupped boards should be carried out as follows:

   (a) Always keep the round side down. This will allow the cut pieces to fall away from the blade and prevent binding (**Fig. 1.21**). *Note* It is important that an even and constant feeding pressure is exerted at all times to prevent any wobble on the round.

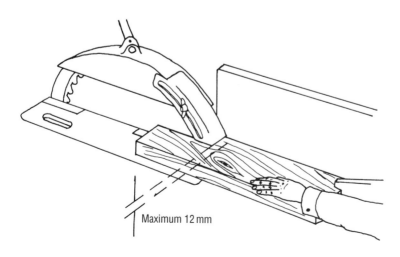

Maximum 12 mm

Fig. 1.18 Setting the crown and extension guards.

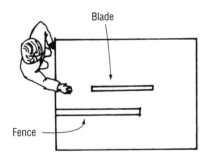

Fig. 1.19 Where to stand when feeding.

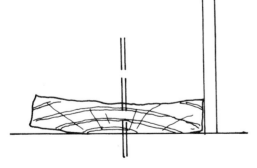

Fig. 1.20 Typical push stick.

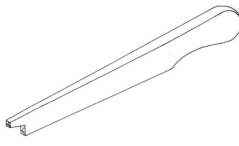

Fig. 1.21 Flat cutting.

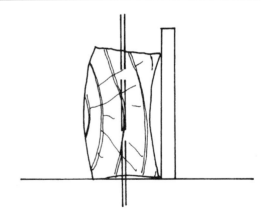

Fig. 1.22 Deep cutting.

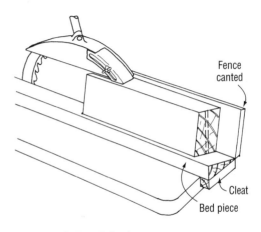

Fig. 1.23 Bevel ripping.

(b) When deep cutting, keep the hollow or cupped side against the fence. This will provide stability as the two points locate on to the fence (**Fig. 1.22**).

4. Secondary functions are often carried out, usually involving some type of saddle or jig. The most common are bevel ripping (**Fig. 1.23**) and angle ripping (**Fig. 1.24**).

1.2.3 Firring strips

These are tapered strips of timber used to produce a slight fall or slope on flat

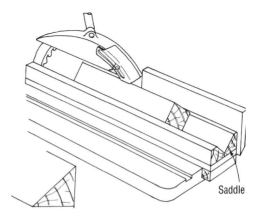

Fig. 1.24 Angle ripping.

roofs, to permit the flow of rain water off the roof members. It must be appreciated that there are several ways of cutting firrings and the following example is the author's chosen method.

1. Mark out the taper as required on to one of the timber pieces. Note that the timber must be at least as wide as one narrow end and one wide end of a firring plus the saw kerf (**Fig. 1.25**).

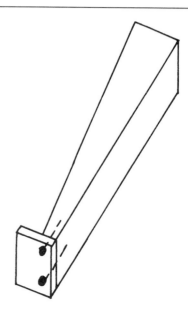

Fig. 1.26 Cleat tacked to narrow end.

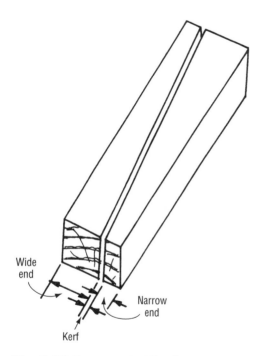

Fig. 1.25 Cut to marked line free hand.

2. Cut to the marking-out line. This must be done free hand on either the rip saw or a narrow band saw.
3. Tack a cleat on to the narrow end (**Fig. 1.26**).
4. Place the timber into the saddle and proceed to cut firrings (**Fig. 1.27**). Wider boards enable more than two firrings to be cut.

1.2.4 Wedge cutting

Figure 1.28 shows a push spike being used. This allows operators to have a

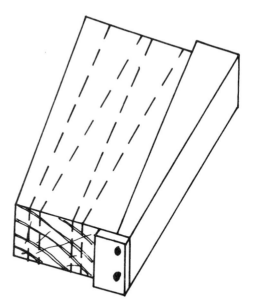

Fig. 1.27 Timber positioned in saddle.

good grip on the material without having their fingers in dangerous areas. It can easily be made by putting a nail into the end of a push stick and filing it to a point. Care must be taken to prevent the nail making contact with the blade.

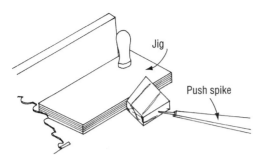

Fig. 1.28 Wedge cutting.

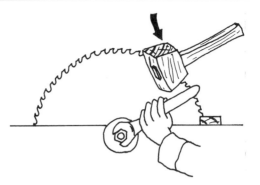

Fig. 1.29 Loosening saw blade.

1.2.5 Changing blades

1. Before fitting a new blade make sure it is the right size. This can be checked by referring to the notice fixed to the machine. This notice, which must be present by law, will state the minimum size of blade allowed on that machine. The largest can be calculated by using the formula

minimum blade diameter =
maximum blade diameter × 6/10

For example, if 360 mm is the minimum blade allowed, divide it by 6:

360 ÷ 6 = 60

and multiply this figure by 10:

60 × 10 = 600

Therefore, the maximum blade diameter allowed would be 600 mm.

2. Isolate the machine.
3. Remove the guard, mouth piece, packings and finger plate.
4. Secure the blade with a piece of scrap timber under the front of the teeth and, by giving sharp mallet blows on the spanner, loosen the locking nut (**Fig. 1.29**). Most machines have left-handed threads that will require loosening by turning the nut clockwise. As a general rule, loosen the nut in the same direction as the blade spins.
5. Remove the clamping collar or flange and blade and put it in a safe place.
6. Fit the new blade on to the spindle shaft and engage the drive pin into the hole provided on the blade, making sure that the pin is at the top.
7. Position the flange on the spindle and secure the locking nut on to the spindle threads.
8. With the pin still uppermost, draw back on the pin with the blade. This ensures that the blade is mounted centrally each time it is replaced. It also prevents the blade moving when contact is made with the material as it spins.
9. Tighten the nut using the mallet to tap the spanner.
10. Reposition the guards, mouth piece, finger plate and packings.
11. Pump start the machine.

1.3 Cross-cut saw construction

1.3.1 Basic functions

Primarily a cross-cut saw is used to:

- cut timber to length;
- square the ends of timber;
- cut out defects such as knots.

They are capable of many more tasks, some of which are covered in this chapter, but it seems very unpopular these days to have such a machine set up for complex operations when a more suitable machine could do the job in a fraction of the time and usually more safely. Work that can be carried out on a cross-cut machine includes trenching, grooving (with the grain), ripping (both straight and bevelled), rebating and even moulding (**Fig. 1.30**). All these operations require the machine to be set up differently to cross-

cutting and in most cases will also require tooling changes.

1.3.2 Design and layout

Travelling head cross-cut

The travelling head cross-cut saw is a very robust machine mainly associated with joinery production in large quantities (**Fig. 1.31**). The saw unit is mounted on to a carriage that is pulled along accurately made steel tracks on hardened steel rollers. This track is supported on a heavy pillar fitted into a cast iron sleeve. The sleeve is bolted to the floor and forms the main body of the machine. To adjust the height of the saw blade a hand

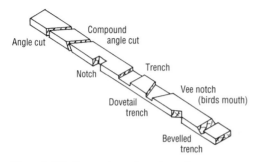

Fig. 1.30 Secondary functions in common use.

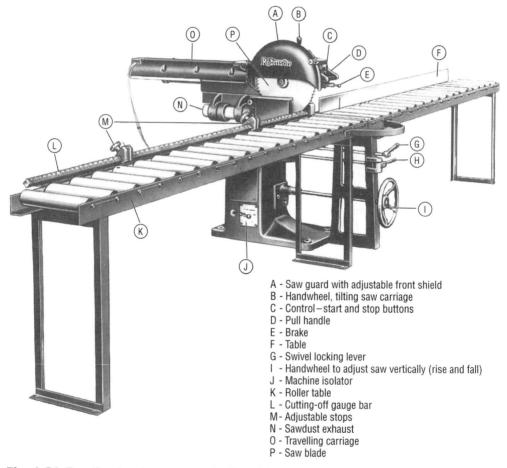

A - Saw guard with adjustable front shield
B - Handwheel, tilting saw carriage
C - Control – start and stop buttons
D - Pull handle
E - Brake
F - Table
G - Swivel locking lever
I - Handwheel to adjust saw vertically (rise and fall)
J - Machine isolator
K - Roller table
L - Cutting-off gauge bar
M - Adjustable stops
N - Sawdust exhaust
O - Travelling carriage
P - Saw blade

Fig. 1.31 Travelling head cross-cut and roller table.

wheel at the front of the machine is turned which, through splayed gears and a lead screw, will raise or lower the pillar inside the sleeve. A locking lever is positioned above the hand wheel and must be slackened off before making adjustments. More importantly, the lock must be engaged after adjustments for two reasons:

- to prevent the pillar dropping through vibration, caused while the machine is running; and
- to prevent the pillar swivelling or turning in the sleeve. This swivelling of the pillar is used to position the head when cutting angles.

The travelling head machine is usually used in conjunction with a roller table which must be set accurately at 90° to the saw blade. On to the table several kinds of stop can be used to ensure and maintain component length (**Fig. 1.32**).

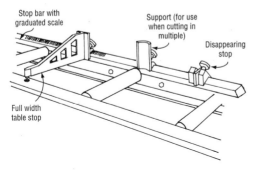

Fig. 1.32 *Types of stop for cross-cut machine fences.*

Radial arm cross-cut

The radial arm cross-cut saw is a much more versatile machine, but of lighter construction (**Fig. 1.33**). The saw unit moves along a cantilever arm from a support column, and can swivel up to 45° in each direction. The saw unit will also cant, allowing angles to be cut in two directions at the same time (compound angles).

1.3.3 Cross-cut saw blades

There are two main types of blade used on cross-cut machines.

a) Parallel plate blades

These can be either alloy steel (**Fig. 1.34**) or tungsten carbide tipped (**Fig. 1.35**). Alloy steel blades are fast becoming obsolete because (1) each tooth must be individually set to give clearance to the body of the blade and (2) the cutting edge quickly becomes blunt when cutting dense abrasive and sheet materials.

Tungsten carbide tipped blades are the most common replacement. This type of blade has a very long life thanks to the hardness of tungsten carbide and therefore withstands the blunting effect of most materials for a far greater length of time.

b) Hollow ground blades

This type of blade requires no set because it tapers towards the centre so that there are no setting inaccuracies, which also results in a better finish (**Fig. 1.36**).

1.3.4 Teeth

The teeth for cross-cutting need to have very sharp points to cut cleanly through the fibres of the wood (across the grain). This is achieved by bevelling the top of the tooth and also applying a bevel to the face of each tooth. The bevels are applied to every tooth but in alternate directions. This allows half of the teeth to cut to the left while the others cut to the right (**Fig. 1.37**).

It is also important to use a negative angle of hook, allowing the teeth points to trail into the cut (**Fig. 1.38**). This produces a scraping cut which, if it were not for the extremely sharp points, would leave a woolly or whiskery finish as the fibres of the wood stand up. The waste produced from cross-cut teeth is very powdery due to this scraping action. Try

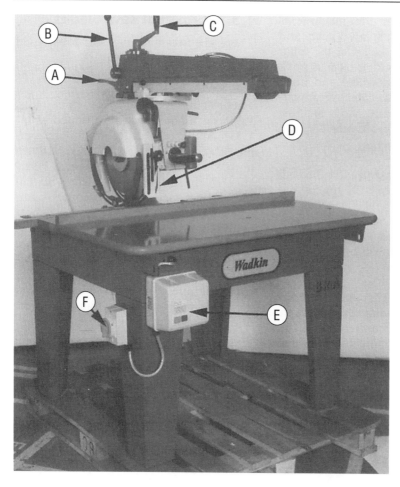

A - Arm locating latch
B - Arm locking lever
C - Pillar rise and fall handle
D - Saw guard visor
E - Start-stop
F - Emergency shut-off

Fig. 1.33 Radial arm cross-cut machine.

to relate the action of these teeth to that of a cabinet maker's scraper.

Note If a positive hook angle is used for cross-cutting, the teeth will dig into the wood because of their hooked shape (or chiselling action) and cause the blade to drag itself uncontrolled across the material. The speed at which it will travel will increase until the saw blade jams and stops in the material or until it reaches the end of its slide. When this happens it is very frightening for inexperienced or unaware operators, not to mention dangerous. The outside edge of a circular saw is usually revolving at about 50 meters per second (mps). A saw unit running out of control could, in theory, reach this speed.

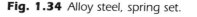

Fig. 1.34 Alloy steel, spring set.

Fig. 1.35 Tungsten carbide tipped.

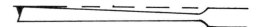

Fig. 1.36 Hollow ground blade.

a) Hardwood cross-cut teeth

The long heel gives extra strength which is needed when cutting hardwoods to prevent the tooth points breaking off.

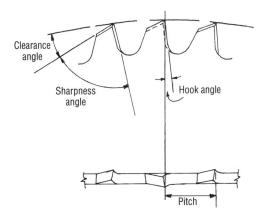

Fig. 1.37 Teeth used for cross-cutting.

(a)   (b)

Fig. 1.38 Cutting action of cross-cut and ripping teeth. (a) Rip saw teeth points dig into the timber before the tooth face is in line. (b) The face of the tooth is always in front of the point to prevent snatching.

Sacrificed for this extra strength is the sharpness of the teeth. The hook angle for this type of tooth is approximately −5° (**Fig. 1.39**).

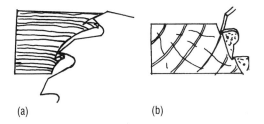

Fig. 1.39 Hardwood cross-cut teeth.

b) Softwood cross-cut teeth

This has a much weaker tooth, with extremely sharp (needle) points and maximum clearance giving a good sawn finish. The hook angle for this type of tooth is approximately −10° (**Fig. 1.40**).

The blades shown are alloy steel but a

Fig. 1.40 Softwood cross-cut teeth.

TCT blade would normally be used. All the angles would be the same and the only difference would be a small tungsten carbide tip on the outer edge.

1.3.5 Trenching

The most common attachments used on a cross-cut machine are the expanding trenching heads and the dado set (also known as gang saws).

a) Expanding trenching heads

These comprise two halves, each with two spur cutters for scoring the shoulders of the cut and two flat cutters for removing the bulk of waste (**Fig. 1.41**).

The width of trench is adjusted by one of two methods, depending on the manufacturer.

- Spacers or shims of varying thickness are inserted between the two halves of the unit and are squeezed together by the spindle locking nut (**Fig. 1.42**).
- Adjustment is by tightening a securing bolt on to a key, set into the shaft, when the required width is

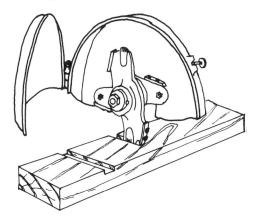

Fig. 1.41 Typical trenching heads on machine.

Fig. 1.42 *Assorted thickness spacers.*

established (**Fig. 1.43**). The lock is very effective and can be run without a spindle locking nut.

b) Dado set

These comprise two outer saw blades to cut the shoulders, with a series of inside cutters built up between them to achieve the required width of trench (**Fig. 1.44**).

It is important, when trenching is being carried out, that a false or auxiliary table is clamped to the machine bed. This will prevent contact with any steel parts, such as the fence or stops, as the head is being pulled into the work (**Fig. 1.45**).

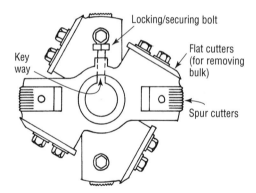

Fig. 1.43 *Trenching heads with securing bolt.*

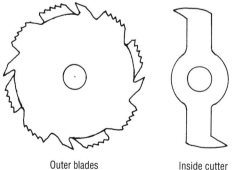

Fig. 1.44 *Dado set.*

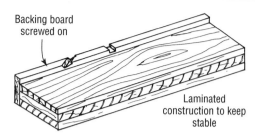

Fig. 1.45 *Auxiliary table for use when trenching.*

1.4 Using a cross-cut saw

1.4.1 Changing the blade

1. Isolate the machine.
2. Raise the blade so that it clears the height of the bed.
3. Pull out the saw blade to a comfortable working position and lock into place with the stops provided. On a machine with no stops a piece of timber can be placed up against the fence and the blade eased back on to it.
4. Remove or adjust guards to allow access to the blade.
5. Place a spanner on to the locking nut and tap it lightly with a mallet until it becomes loose. Remember, the nut is usually designed to tighten itself as the machine is started by having a left-handed thread.
6. Remove the packing collar and outer flange.
7. Take off the blade and fit the new one on to spindle shaft, making sure that the drive pin is at the top.
8. Refit the outer flange and packing collar.
9. Tighten the nut finger tight and draw back on the blade to pull it up against the pin.
10. Tighten with a spanner and mallet, taking care not to overtighten.
11. Refit guards to comply with current regulations and revolve blade by

hand to check that it does not rub or catch on anything.

12. Pump start the machine (start and stop to allow the speed to build up slowly, listening for any rubbing or catching).

1.4.2 Cutting bowed boards

a) Face bowed

Place the round side down to the bed to allow the cut pieces to fall away from the saw blade on completion of the cut thus preventing them from binding (**Fig. 1.46**).

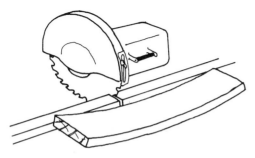

Fig. 1.46 Cutting face bowed boards.

b) Edge bowed

Still following the rule for face bowed, keep the round edge to the fence with the peak of the curve at the point where the blade enters. This will prevent the timber from being pulled violently back to the fence once cut (**Fig. 1.47**).

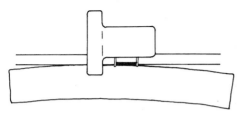

Fig. 1.47 Cutting edge bowed boards.

1.4.3 Setting up for trenching

Using keyed units with securing bolts:
1–6. Follow steps 1–6 of Section 1.4.1 (changing the blade).

7. Remove the blade and unscrew the drive pin from the inner flange.
8. Fit the purpose-made key into the groove in the spindle shaft and slide the first half of the trenching head on to the spindle.
9. Turn the securing bolt on the trenching head so that it grips firmly on to the key.
10. Slide the second half of the trenching head into position and check for accuracy of width. This is done by turning the unit by hand to allow both scribing cutters to scratch the surface of a scrap piece of timber and then measuring the distance between them (**Fig. 1.48**). When the width of cut is correct secure into place with the bolt.

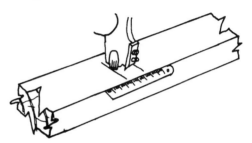

Fig. 1.48 Checking width of cut.

11. Position the auxiliary table on to the machine bed and secure with G cramps. A false table is not only used to prevent the blades being pulled into the metal fences but also to act as a backing board. This will give support to the material being cut and prevent any spelching or break-out of the fibres of the wood, resulting in a good finish. The auxiliary table is also useful for lining up marking-out lines to the point at which the blades will enter, thus ensuring accuracy.
12. Refit the guards to comply with current regulations and spin the heads by hand to ensure that nothing catches.
13. The only job remaining is to set the depth of cut. An experienced

machinist will do this by trial and error. If the first guess is wrong, adjustments will be made by turning the hand wheel at the front of the machine. This is repeated until the correct depth is achieved. For beginners it is a good idea to mark the pillar with a pencil and measure the amount of movement against this pencil line and the cast iron sleeve.

1.4.4 Bevelled and angled work

Depending on the type of machine and manufacturer, different adjusting methods will be incorporated. Whichever type is used a scale will be present to set to the number of degrees accurately. **Figures 1.49–1.51** show a cross-cut machine set up to carry out a few secondary functions.

1.4.5 Alternatives to a standard machine

Hydraulically operated machines are commonly used in industry. They are operated by a foot pedal which when

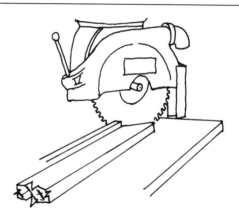

Fig. 1.50 Angle cutting (mitres).

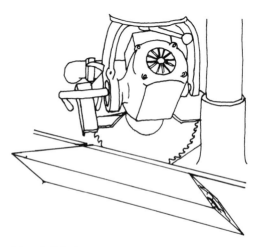

Fig. 1.51 Compound angle cutting.

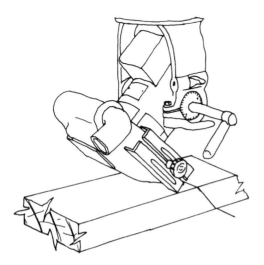

Fig. 1.49 Bevel cutting.

depressed activates the hydraulic mechanism bringing the saw across the table. It can also be linked up to a roller table so that the rollers will turn by moving a pedal or lever to feed the timber up to the required stop.

Computer numerically controlled (CNC) machines are also available. When programmed they will measure a long length of timber by means of an arrangement of electronic eyes and cut out short lengths as required. They are very expensive to buy but, due to their speed and accuracy, once programmed they prove to be much more efficient on large batches or in mass production.

1.5 Dimension saw construction

1.5.1 Design and layout

The dimension saw is the most versatile of all circular sawing machines available. Its primary functions are cross-cutting and ripping (**Fig. 1.52**).

The machine is similar in design and layout to a circular rip sawing machine but two major differences distinguish the two:

- a sliding table, which can be locked into place when ripping and also has a graduated scale of angles for use when cross-cutting;

- a canting saw unit for bevel ripping and compound angle cross-cutting.

The dimension saw is designed for precision work rather than high levels of productivity as it can only incorporate blades of 450 mm diameter and less. Although ripping is often carried out on the dimension saw, and there is nothing wrong in doing this, it is more common to use a circular rip saw. The reason is that because dimension saws are used for cross-cutting as well as ripping they are fitted with a general-purpose blade that will not do the job as efficiently. These general purpose blades are particularly useful when cutting plywoods because of

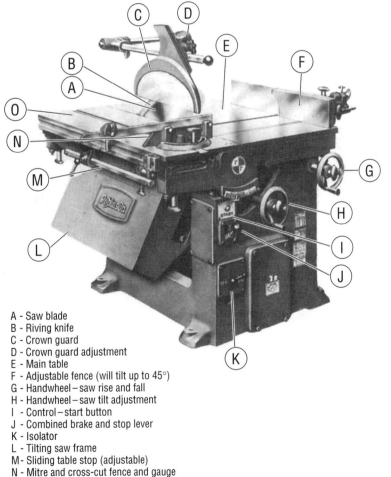

A - Saw blade
B - Riving knife
C - Crown guard
D - Crown guard adjustment
E - Main table
F - Adjustable fence (will tilt up to 45°)
G - Handwheel – saw rise and fall
H - Handwheel – saw tilt adjustment
I - Control – start button
J - Combined brake and stop lever
K - Isolator
L - Tilting saw frame
M - Sliding table stop (adjustable)
N - Mitre and cross-cut fence and gauge
0 - Sliding table (rolling)

Fig. 1.52 Dimension saw machine.

the opposite grain directions. Because the teeth are usually TCT, the abrasive characteristics of plywood and all other manufactured sheet materials will not have too much of a blunting effect on the teeth. Sheet materials with plastic laminated surfaces should be cut with the laminate facing up. This will prevent any spelching (chipping) as the saw passes through.

Where large quantities of sheet material are to be processed, a range of machines are available that can cope more easily with the large sheet sizes. They are known as panel saws which are basically dimension saws with some additional features (**Fig. 1.53**).

- The fence is mounted on to a graduated bar which extends back much further than a normal saw, so that a full sheet can be trimmed.
- A pivoting extension table on the opposite side allows sheets to be supported when ripping and cross-cutting.
- There is no supporting arm (pillar) for the crown guard. Instead it is fitted on to the riving knife thereby allowing

sheets to pass through the machine at any width of cut without being obstructed. Some machines even have the crown guard acting as a top pressure by the addition of a special roller guard. As the sheet is brought forward the guard will spring up and apply pressure throughout the cut. On completion of the cut the guard will return to the bed and completely enclose the blade.

- A scoring saw can be fitted in front of the main blade to precut the underside of the sheet (possibly wood veneered or plastic laminated) thereby preventing any break-out. The scorer will be running in the opposite direction to the main blade and will be set to cut approximately 3 mm into the material.

1.5.2 Sawtable

The table is made up of two sections (**Fig. 1.54**). The right-hand section (A) is fixed with a ripping fence bolted on to it by means of a tee slot. With the tee slot undone the fence can be adjusted to give

Sliding table   Extension table

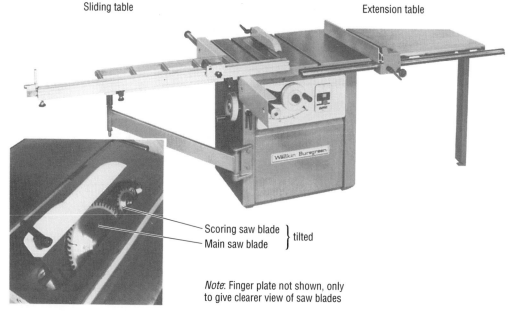

Scoring saw blade } tilted
Main saw blade

Note: Finger plate not shown, only to give clearer view of saw blades

Fig. 1.53 *Typical panel saw and blade arrangement.*

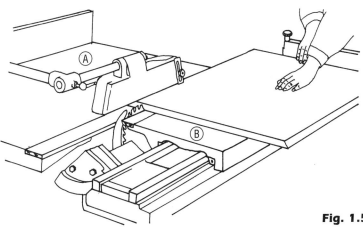

Fig. 1.54 Table arrangements.

a width of cut. An extended bed allows the fence to be moved further away for larger components such as panels and sheet materials.

The left-hand side (B) is a sliding table that runs on rollers for use when cross-cutting. When ripping, this table can be locked in place by the use of a spring-loaded pin underneath it. For changing the blade side locks will release the table, allowing it to be pulled out sideways to give the operator space to work.

For cross-cutting, an adjustable mitre fence can be secured on to the sliding table by a common pivot joint and locking nut which will allow angles to be cut in both directions of varying degrees (**Fig. 1.55**).

1.5.3 Saw unit

The motor is integral with the saw spindle (direct drive) and can be canted (or angled) up to 45° via a hand wheel and set accurately from a quadrant plate and finger, found at the front of the machine (**Fig. 1.56**). The extraction hood is part of the moveable unit, thus preventing any problems when canting or altering the height of the saw.

Because the blade can be canted, it eliminates the need for jigs on certain operations. However some jobs, even on a machine as versatile as a dimension saw, have to be cut with the aid of a jig.

A good example is a taper for a table leg (**Fig. 1.57**).

1.5.4 Brake

Dimension saws are usually fitted with a brake that can also be used as the stop button (**Fig. 1.58**). This means that when the brake is applied the machine is automatically switched off. Ideally the position for this brake is at either knee height or as a foot pedal so that the operator can stop and brake the machine without removing their hands from the material being cut.

1.5.5 Fence

As with all circular sawing machines the fence is set with a lead. For example, if a 600 mm saw blade is being used the fence should be running away from it by 0.3 mm from front to back (see **Fig. 1.15**). Adjusting screws are set into the fence to alter the amount of lead, but it is unlikely that it will need altering once it has been set.

The reason for this lead is to prevent the timber binding as it is cut due to the fence pressing it against the blade. Failure to do so will create a number of problems.

- Because the timber is still touching the blade it will create thermal stress. This is simply a build-up of heat

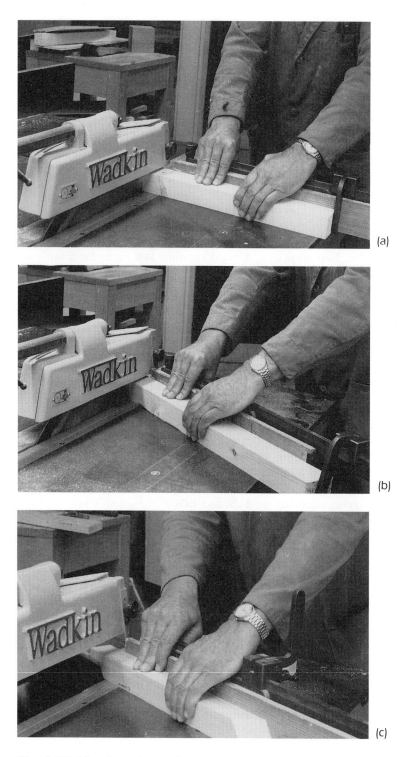

Fig. 1.55 Mitre fence in use: (a) square cut; (b) angle cut; (c) compound angle cut.

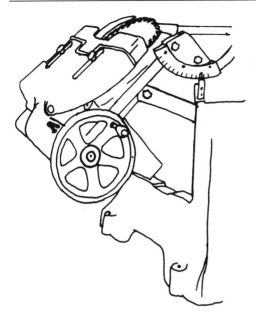

Fig. 1.56 Canting unit.

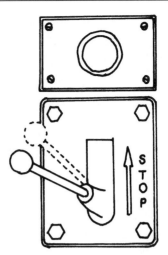

Fig. 1.58 Combined brake and stop lever.

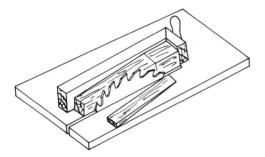

Fig. 1.57 Jig for tapering.

caused by friction as the timber rubs on the blade.

- A series of score marks will be present on the cut surface. These are caused by the back of the blade (up-running teeth) and can sometimes be used as a tell-tale sign to let the operator know that more lead is required.

1.5.6 Dimension saw blades

Very few dimension saw machines, if any, will be working with alloy steel blades. TCT will be the most common and some, particularly the larger companies who process large quantities of sheet material, will be working on poly crystalline diamond (PCD or often referred to simply as 'diamond') tooling. This is the latest development in hard wearing, low maintenance tooling. It is more expensive than TCT but will last much longer. A suitable hook angle for a dimension saw, regardless of the type of blade, would be about 5–10°.

Self-assessment

The following questions have been written around the previous text in this chapter. If you cannot answer any of the questions, simply restudy the respective areas. Good luck!

1. What are packings used for on a circular rip saw?
   (A) fine adjusting the fence position
   (B) to keep the blade running true
   (C) to act as a guard to the back of the blade
   (D) adjusting the blade height.

2. What is the most suitable blade for cutting hardwood?
   (A) needle point
   (B) alloy steel
   (C) TCT
   (D) any blade with 30° of hook.

3. Which of the following is the correct position for a fence when ripping timber?
   (A) base of fence level with the gullets of the blade
   (B) base of fence level with the riving knife
   (C) base of fence level with the crown guard
   (D) base of fence level with the front of the blade.

4. What is the maximum height that a crown guard can be above the material being cut?
   (A) 6mm
   (B) 10mm
   (C) 12mm
   (D) 15.5mm.

5. When should a push stick be used for feeding a piece of timber 1.2m long on a circular rip saw?
   (A) the last 450mm
   (B) all the way through
   (C) the last 300mm
   (D) only for removing the off-cuts.

6. Explain in your own words why it is necessary to draw a blade back on to the drive pin when refitting it to a rip saw bench.

7. Produce a line diagram showing a rip saw tooth suitable for ripping softwood and indicate the correct hook angle in degrees.

8. Explain with the aid of diagrams how the following boards should be fed:
   (A) face-bowed board on a cross-cut machine
   (B) cupped board being flat cut on a circular rip saw machine.

9. Dado sets are an efficient way to cut a trench across a piece of timber on the cross-cut machine, but how is the width of cut altered?

10. Explain the two main differences between a circular rip saw machine and a dimension saw machine.

Crossword puzzle
Circular sawing machines

From the clues below fill in the crossword puzzle. The answers could be related to the circular rip saw, the cross-cut saw or the dimension saw.

Across

1. This is set with a lead from the blade.
3. Used to feed the last 300mm of a cut.
6. Wheel-like device used to assist the feeding of timber on a cross-cut bed.
7. Found on trenching heads – used to cut the shoulder.
8. Metal tips on a saw blade.
10. Cross-cut teeth for cutting soft-woods.
14. ----- steel, used to make spring set blades.
15. Blade has this applied when bevel ripping on a dimension saw.
16. Type of sawing machine.
18. ----- over, type of cross-cut saw.
20. Not sharp.

21. Stops a motor much quicker.
22. --------- table for use when trenching.

Down

1. Pressed between the blade and finger plate.
2. Safety device above saw blade.
4. Type of hat, also found on a tooth.
5. 1200mm from the back of the blade to its end.
7. Used to maintain component lengths.
9. Bending of teeth to give clearance.
11. Type of trenching head.
12. Groove across the grain.
13. Not a pull over one.
15. To rip boards into smaller sizes.
17. Rise and ----.
19. Width of saw blade cut.

2 Bandsawing machines

2.1 Band resaw

2.1.1 Functions

The band resaw is used to convert (rip) timber as with a circular rip saw but due to the size of this type of machine and the ease with which it cuts, it has certain advantages over circular sawing machines:

- a much faster feed rate;
- smaller saw kerf so less waste;
- deeper cuts are made easier (deep cutting).

2.1.2 Design and layout

The machine bed must be at a comfortable height for the operator to work safely. To allow this, due to the size of the pulley wheels (36 in diameter being the most common size), a pit or hole must be cut into the floor so that the bottom wheel can be lowered below ground level (**Fig. 2.1**). The machine's extraction pipe is usually situated in this pit to aid the removal of the waste. All resaws are fitted with power-driven rollers to feed the material past the blade. The position of the rollers is adjusted by a hand wheel at the front of the machine. For more immediate response a foot pedal at the front can be pressed to remove roller pressure.

2.1.3 Saw wheels

Saw wheels on which the blade runs can be crowned or flat on the face and are always balanced. The top wheel can be adjusted vertically by hand wheel for the application of blade tension (or saw straining). On older machines this straining is done with weights on a lever arm, on more modern machines by coil spring. When the blade is tracked (the tilting of the top wheel to keep the blade in the correct position) it must be realized that the teeth on the blade overhang the wheel to prevent damage to both the wheels and the teeth (**Fig. 2.2**).

2.1.4 Machine table (bed)

The bed is fitted with sectional anti-friction rollers at both the infeed and outfeed ends. They are accurately set by the manufacturer to be slightly higher than the bed itself so that the timber will feed through more easily (**Fig. 2.3**).

2.1.5 Fence

The fence is easily adjusted to give the desired width of cut and is usually extremely accurate. Adjustments are made by turning the hand wheel and reading off a scale to give the desired position. Once set, a locking lever is used to set the position. (*Note* Always make sure that the

Fig. 2.1 Band resaw layout.

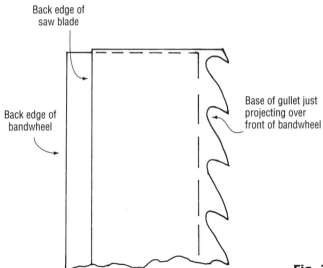

Back edge of
saw blade

Back edge of
bandwheel

Base of gullet just
projecting over
front of bandwheel

Fig. 2.2 Flat wheel and blade position.

locking lever is fully engaged. Failure to do this could allow the fence to wander. This happens when the feed rollers are pressing the material against the fence.) The fence, just like the bed, is fitted with antifriction rollers to aid feeding (**Fig. 2.3**).

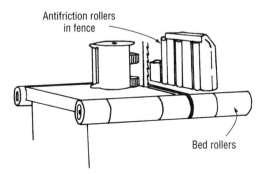

Antifriction rollers
in fence

Bed rollers

Fig. 2.3 Antifriction rollers in the bed and fence.

If angled cutting is required the fence can be canted and locked at the desired angle. It is not very common, however, to have a calibrated scale for canting. Usual practice is to set a sliding bevel to the angle required and then place this on to the bed and adjust the fence until it matches the bevel (**Fig. 2.4**).

2.1.6 Feed rollers

Two feed rollers are mounted on to a shaft rising up through the bed. They are driven by an independent electric motor which can only be started when the blade is running at the optimum running speed. This will prevent the timber being fed into the blade before it can cut efficiently and therefore prevent the blade being pushed off its wheels. The standard rollers are made of steel and have a serrated finish to give a good grip on the timber (**Fig. 2.5**). On most machines the rollers are vertically adjustable on the shaft so that support can be given to deep sections of timber at the top and

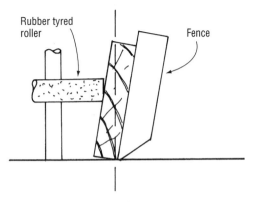

Rubber tyred
roller

Fence

Fig. 2.4 Angling the fence.

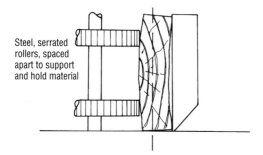

Steel, serrated rollers, spaced apart to support and hold material

Fig. 2.5 *Rollers set for deep cutting.*

bottom, thereby allowing greater accuracy of cut.

The steel rollers can be removed and replaced with rubber coated rollers for cutting preplaned timber or bevelled boards to prevent the serrations marking the face (as in **Fig. 2.4**).

The rollers are spring loaded to allow for a variation in stock size. For controlled feeding the foot pedal should be pressed to allow the first piece to enter without the rollers having to spring out of the way. This will prevent the timber from 'kicking' as the rollers press on to it (**Fig. 2.6**). Once feeding has commenced with the first piece, the pedal can be left alone while following piece are butted up to the one previous. For the last piece the pedal should be 'held' (by foot) to stop

the rollers from jumping in when the cut is complete and the timber has gone.

2.1.7 Feed speeds

Most machines have means of adjusting the feed speed, usually ranging from 4 metres per second (for cutting deep sections) to 60 mps (for cutting thin softwood).

Many factors will affect the choice of feed speed, such as type of timber, moisture content and timber spices. It is also necessary to consider the type and design of tooth on the blade. Just like cross-cut and rip sawing machines, blade tooth design will vary depending on the type of work being done.

Gullet area is one of the most important factors to consider (**Figs 2.7** and **2.8**). This is because, if the gullet is too small in relation to the feed speed, it will not be capable of removing and expelling the saw dust, thus causing the blade to clog up. This could in turn lead to the blade being pushed back on the wheels, causing untold damage. To overcome this problem the feed speed must be reduced, allowing the blade to cut efficiently. Alternatively, if a fast feed speed is essen-

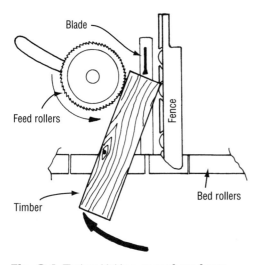

Fig. 2.6 *Timber kicking away from fence.*

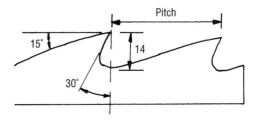

Fig 2.7 *Standard softwood teeth.*

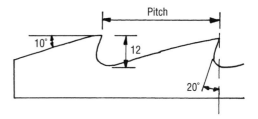

Fig. 2.8 *Standard hardwood teeth.*

tial, a blade with a larger gullet area must be used.

On the opposite side of the scales, when feeding slowly, the blade can surge forward on the wheels. This is caused by too much hook angle, which allows the teeth to drag themselves forward or snatch into the timber. This problem can be partially overcome by increasing the feed rate but it is advisable to change the blade for a more suitable one.

2.1.8 Set

Most resaw blades are swage set. This is where the points of each tooth are squashed out to be wider than the body of the blade. To keep the set on all the teeth the same they are dressed with a dressing tool. A dressing tool is basically two steel rollers clamped together with a gap between them for the blade to fit in. It is worked similar to a hand plane, the two handles firmly gripped and then worked over all the teeth. For cutting abrasive timbers it is recommended that **stellite** tipped blades be used as they are far more hard wearing (similar to TCT.) Also, when cutting abrasive timbers, a fast feed is advised to prevent the teeth from rubbing on the material and therefore increasing the working life of the blade.

2.1.9 Cleaning devices

In order that the dust and resin does not build up on the blade and wheel faces, cleaning devices are fitted. Machines are sometimes supplied with only one device which simply keeps the blade clear, the idea being that if the blade is free of this build-up it cannot be transmitted to the wheel faces. In fact the best arrangement is to have devices for both the blade and wheels. Wheels are kept clean by the use of spring-loaded or counterbalanced scraper plates (**Fig. 2.9**).

For keeping the build-up on the blade to a minimum, spring-loaded felt pads

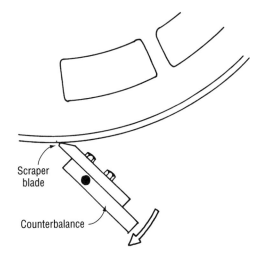

Fig. 2.9 Counterbalanced scraper.

are utilized that lightly nip the blade between them as it travels during cutting. To assist the felt pads an adjustable drip feed releases a flow of cleaning fluid (paraffin mixed with oil, 50:50 is usual) on to the pads (**Fig. 2.10**). They will soak up the fluid and smear it on to the blade every time it passes between them. This not only helps keep the blade clean but also the wheels, by the fluid passing from the blade on to them. It is good practice to clean these cleaning devices from time to time and free them of any debris that

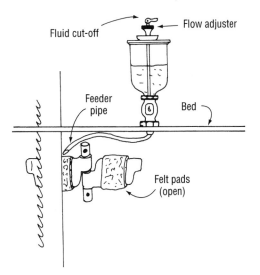

Fig. 2.10 Fluid feed.

may have built up. Note that the teeth are not to be touching either of the cleaning devices. The felt pads would simply be chewed up as they pass through. This means that the teeth will sometimes have resin deposited on them. Again, it is worth cleaning this resin off periodically to allow the blade to cut efficiently.

To prevent splinters dropping between the blade and bottom wheel a mouthpiece is set into the bed. Just like on a circular rip saw, several sizes may be needed to facilitate different blade sizes.

2.1.10 Saw guides

All machine are fitted with guides to keep the blade running in a true and accurate line parallel to the fence (**Fig. 2.11**). They must be checked and, if needed, adjusted to be as close to the blade as possible. They are mounted on to a height-adjustable arm which must be set close to the work piece so that the guides are effective near the part of the blade which is doing the work.

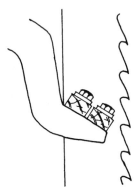

Fig. 2.11 *Saw guides.*

2.1.11 Guards

This type of machine is usually well guarded, not only by the manufacturers' standard guards but also by a sound enclosure. Owing to the high noise output that this type of machine generates, enclosures are often found. Both the enclosure and the guards are easily removed or swung away to allow access for maintenance and blade changes.

2.2 Narrow bandsaw construction

2.2.1 Functions

The basic functions of a narrow bandsaw are quite simply 'cutting'. This term can then be divided into two main categories:

- straight – either free-hand or with the aid of a fence;
- curved – either free-hand (working to a pencil line) or with jigs/templates for repetition work.

In reality a narrow band saw machine is used for probably hundreds of separate tasks from complicated curve-on-curve work (look at a banana to see curve-on-curve) to the simplest of jobs such as cutting haunches on to a tenon. It may sound contradictory, but one disadvantage of the bandsaw is that it is so easy to use. If a machine is noisy people are straight away put off it and this awareness aids concentration. If a machine is complex the operator is always thinking about it while working. The bandsaw is neither, and for this reason it is responsible for a large percentage of accidents in wood machining workshops.

2.2.2 Design and layout

A bandsaw is basically a continuous steel band with teeth running on one edge. This band is placed on two pulley wheels, the bottom one being driven by a motor. The wheels are coated with hard rubber tyres to prevent the teeth of the blade and the wheels becoming damaged while running. The top wheel is suspended from a swan-neck casting that allows room for a large working area (**Fig. 2.12**).

Fig. 2.12 Narrow bandsaw.

2.2.3 Tension (or strain)

The top wheel can be raised or lowered in order to stretch the blade around the wheels. This is known as applying tension or straining the blade. Tension is very important as without it, or even with the wrong amount, blades can be pushed off the pulley wheels when timber is fed in to be cut.

Before tension can be applied a tension gauge must be set to suit the width of blade being used (**Fig. 2.13**). This gauge determines the amount of pull or force to obtain the correct blade tension. It must be realized that a narrow blade needs less force than a wide one to keep it tensioned on the wheels.

When the gauge is set the top wheel is raised by turning the hand wheel until the tension indicator reaches the correct position (**Fig. 2.14**).

Some machines incorporate both the tension gauge and indicator on one scale. As the hand wheel is turned to raise the top pulley wheel a pointing device travels along a scale marked with blade sizes.

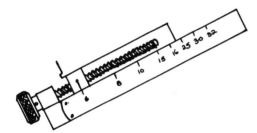

Fig. 2.13 Tension gauge.

Fig. 2.15 Tension test (twisting).

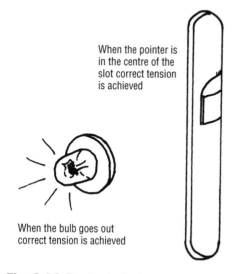

When the pointer is in the centre of the slot correct tension is achieved

When the bulb goes out correct tension is achieved

Fig. 2.14 Tension indicator.

When the pointer registers on to the calibration marked as the blade's width, the setting is finished. It must be remembered that whichever device is used too much strain is just as bad as too little. Overtensioning can simply snap a blade.

Where no gauge or indicator is present the operator's judgement is required to assess the correct tension. As a general rule use the following guides:

1. *45° of twist.* Lightly grip the blade between finger and thumb and gently twist/turn the blade. If you cannot reach 45° then the tension is too much and if the blade passes 45° there is not enough tension (**Fig. 2.15**). Note that if the teeth are digging into your hands then you are using too much force.

2. *9 mm of movement.* Making sure that the guides and thrust wheels are clear of the blade apply pressure as indicated in **Fig. 2.16**. It is difficult to measure, but approximately 9 mm of movement should be possible in each direction.

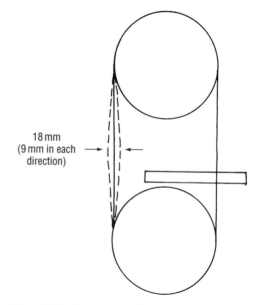

18 mm (9 mm in each direction)

Fig. 2.16 Tension test (9 mm of movement).

Tension on old machine was carried out by weights on a lever arm, but more modern machines use a helical spring. Helical spring tension devices are far superior and even allow a cushion effect should the blade make contact with foreign bodies such as nails and screws.

2.2.4 Tracking

A bandsaw blade will run on any part of the wheels (front, back or centre) but experience will show the centre to be the safest. If positioned at one of the edges, sudden impact or cutting curves too small for the blade could cause it to come off, causing damage to the blade, machine, material being cut and even the operator. To allow the operator to position the blade in the centre each time it is changed, a tracking device (**Fig. 2.17**) is fitted to the machine. By turning the rise and fall hand wheel the top pulley will cant in either direction to track the blade into position.

When fitting a new blade it is necessary to spin the wheel by hand to allow the blade to settle in and find the running position. Each time an adjustment is made the wheel must be turned again. By doing this you will be able to watch the blade and see if it stays in the intended position or moves forward or back. It is better to spend a few extra minutes getting the setting right than

finding that on starting the machine up the blade runs off the wheels.

2.2.5 Thrust wheels and guides

Saw guides are fitted to bandsawing machines to keep the blade running in a straight path. They are ideally made of a hard wearing material such as beech or lignum vitea and positioned as close to the blade as possible without actually touching. They must also be just behind the gullet area of the blade to prevent the teeth from cutting into them.

Where the guides stop the blade moving left to right, a thrust wheel is incorporated to prevent the blade being pushed back off the pulley (**Fig. 2.18**). This consists of a ball racer bearing held in a cast housing. It should be set approximately 1 mm from the back of the blade. A new, sharp, clean blade will probably cut without touching the thrust wheel; it is only when the blade looses some of its efficiency that the thrust wheel will be relied upon to keep the blade in place. A thrust wheel and set of guides are positioned above and below the bed and both must be set the same.

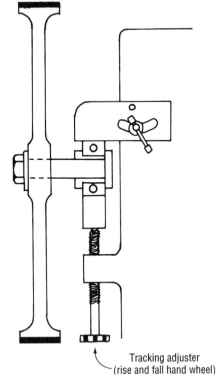

Tracking adjuster (rise and fall hand wheel)

Fig. 2.17 Tracking device.

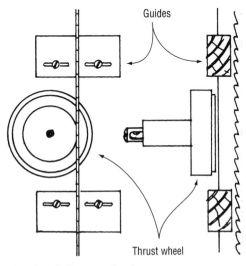

Guides

Thrust wheel

Fig. 2.18 Thrust wheel.

2.3 Using a narrow bandsaw

2.3.1 Secondary functions

The largest proportion of work carried out on a narrow bandsaw is classed as plain cutting (this can be straight or curved). From time to time, however, bevel cutting is required which can be carried out with the aid of jigs, saddles or bed pieces just as on a circular rip saw. Alternatively, the bed can be canted (**Figs 2.19** and **2.20**).

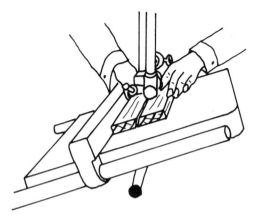

Fig. 2.19 Bevel cutting.

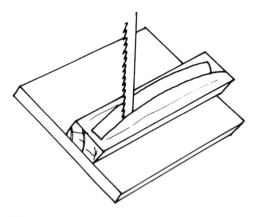

Fig. 2.20 Circular bevel.

2.3.2 Guards

Before any work is started, guards must be set in compliance with current regula-

tions. Basically the requirement is to keep only the cutting part of the blade 'open'. The guard above the bed is to be adjusted as close as possible to the material being cut (**Fig. 2.21**).

2.3.3 Selecting the right blade

Other than poor blade joints, the most common cause of blade breakages is unsuitability for the type of work being done.

Various tooth shapes and designs are available to suit the type and size of material (**Figs 2.22** and **2.23**).

Most factories now use throw-away hard-point blades which are suitable for general work. It is useful, however, to have a few blades with different pitch distances. As a general rule small pitch blades (6 points per inch) are suitable for cutting sheet materials with low moisture/resin contents. It would not take long to clog this blade if it were used on Redwood (Scots pine) and for this material a larger pitch is required (3 or 4 point per inch). Also remember that narrow blades will distort when used for cutting deep sections of material.

Above all, the pitch must not be bigger than the material's thickness (**Fig. 2.24**). If it were the timber could in theory pass into the gullet and generate excessive stress in the blade. Also, the finish would be poor as the teeth smash through large portions of the timber.

When cutting curves, use a blade no bigger than a line drawn inside the curve (**Fig. 2.25**). The important factor is that the back of the blade must not rub on the material as it is being fed.

2.3.4 Fitting a new blade

The following procedure should be observed.

1. Isolate the machine.
2. Remove the tension from the blade by turning the hand wheel.

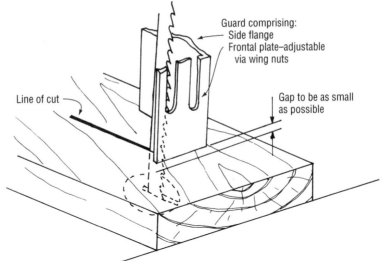

Guard comprising:
Side flange
Frontal plate–adjustable
via wing nuts

Line of cut

Gap to be as small
as possible

Fig. 2.21 Setting the guards.

Fig. 2.22 Standard, general purpose tooth.

Fig. 2.23 Blade for cutting deep sections of softwood.

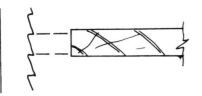

Fig. 2.24 Relationship between pitch and material.

3. Swap the blade for a new one of the required width and check for any defects. Fold the old blade and store safely away.
4. Move the thrust wheels and guides back, well clear of the blade.
5. Set the tension gauge to suit the blade width selected.

6. Place the blade in the centre of the pulley wheels.
7. Apply enough tension to take up the slack in the blade (not the full amount).
8. Rotate the top wheel by hand to allow the blade to settle in and find its running position. If it moves away from the centre of the wheels, adjust the tracking device until the blade returns and stays in the centre.
9. Apply the required amount of tension by consulting the tension indicator or by carrying out the tests.
10. Again rotate the top wheel by hand to check the tracking of the blade with full tension applied.
11. Set the thrust wheels and guides to suit the blade width.

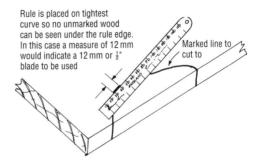

Rule is placed on tightest curve so no unmarked wood can be seen under the rule edge. In this case a measure of 12 mm would indicate a 12 mm or ½" blade to be used

Marked line to cut to

Fig. 2.25 Selecting blades for curved work.

12. Position guards to suit material being cut.
13. Turn on the power and pump start the machine, listening for any tell-tale noises.

2.3.5 Folding blades

In order to keep the blades sharp and in good condition they must be neatly and safely stored. This involves folding the blades into three loops as follows:

1. Hold the blade with palms facing up (**Fig. 2.26a**).
2. Bring wrists around to face each other by twisting. Make sure that points X and Y come out to the front and start to cross over each other.
3. As X and Y pass in the middle let the section of blade between your hands fall backwards, in front of your body (**Fig. 2.26(b)**).
4. Gently lower the blade to the floor (**Fig. 2.26(c)**).

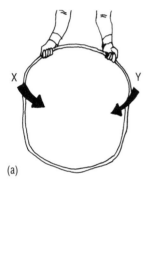

(a)

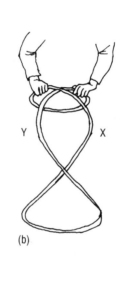

(b)

(c)

Fig. 2.26 Folding a blade into three loops.

Self-assessment

The following questions have been written around the previous text in this chapter. If you cannot answer any of the questions, simply restudy the respective areas. Good luck!

1. What is the function of a thrust wheel?
   (A) for the blade to travel around
   (B) to follow the con tour of a component
   (C) to give support to the blade while cutting
   (D) to track the blade.

2. Why must a blade be tensioned?
   (A) to keep it running in the centre of the wheels
   (B) to keep it running on the front edge of the wheels
   (C) to prevent the guides from wearing out
   (D) to take up the slack and keep it running tightly on the wheels.

3. When selecting a narrow bandsaw blade for straight cutting, which is the most important factor to consider?
   (A) radius of tightest curve to be cut is not bigger than blade width
   (B) pitch length does not exceed the thickness of the material being cut
   (C) whether the wheel is rubber coated or not
   (D) the amount of components to be cut.

4. Why is the bottom wheel of most band resaw machines sunk into a pit in the floor ?
   (A) to allow larger sections of timber to be cut
   (B) to prevent saw dust being blown around the work shop
   (C) to create a safe working height
   (D) for safety reasons – to prevent the operator making contact with the blade.

5. A drip feed supplying paraffin and oil is fitted to resaws to:
   (A) prevent the teeth clogging up with resin
   (B) to lubricate the machine
   (C) to allow the material to be fed easily
   (D) to assist the felt pads in keeping the wheels and the body of the blade clean

6. (A) Explain the considerations to be taken into account when selecting a blade for cutting curved work.
   (B) State and show with a sketch a practical method of making your selection.

7. Explain when it would be necessary to remove the feed rollers on a band resaw and refit rubber-coated ones.

8. Explain the function of saw guides as fitted to both narrow bandsaws and resaws.

9. When feeding the first piece of material into a resaw, how should the rollers be used?

10. Why is the gullet area an important part to consider when selecting a blade?

Crossword puzzle
Bandsawing machines

From the clues below fill in the crossword puzzle. The answers could be related to the band resaw or the narrow bandsaw.

Across

1. (and 3) Device behind the blade to stop it being pushed back.
3. See 1 across.
6. Square or 90°, chamfer or 45° – both are known as ------.
7. Heavy hardwoods are usually this, and abrasive.
8. The clearance length on a tooth can also be this.
11. The beginning! To get the blade moving.
13. Before altering the guards do this to the machine.
14. Used to keep the blade running straight.
17. An oily one is used to wipe the machine.
19. Not a narrow bandsaw.
20. Used to power a motor.
21. What you do to a stop button.
22. Teeth on a feed roller also on a bread knife.

Down

1. Term given to the process by which the blade is positioned centrally on the wheels.
2. This is carried out as a check after changing a blade or before running a job off.
4. Work is completed.
5. A blade's ------ is important to enable it to fit around both wheels.
9. Wrapped around the wheels and covered in teeth.
10. A scraper on a resaw but you also put your dinner on one.
12. A blade must never be started until it has been tracked and ---------.
15. Below the table also where Australia is – down -----.
16. To change – used to isolate a machine.
17. Expensive car, also found on a band saw to cut down on friction.
18. The round in the bottom of a tooth.

3 Planing machines

3.1 Surface planer

3.1.1 Basic functions

Classed as the machine's main functions are:

- planing, or smoothing the surface of, solid timber to be true, flat and out of twist;
- planing two adjacent sides square (90° to each other) before thicknessing to finished dimensions. The accuracy of following cuts (moulding and rebating) is determined by how accurately the face side and face edge have been squared.

3.1.2 Secondary functions

The surface planer is also used a great deal for:

- preparing edges for glue jointing; and
- bevelling, chamfering and rebating, all with the aid of additional safety devices.

3.1.3 Design and layout

The machine consists of a main frame supporting two tables with a cylindrical cutter block placed in between them and a fence on the front table to locate the material against (**Fig. 3.1**).

3.1.4 Tables

These are usually referred to as the front and back tables but are often termed differently from one region to another. Other names include infeed and outfeed beds, front and rear tables. Both tables are independently adjustable horizontally, where after undoing the table locks the whole table will pull out, away from the block. This allows access to the block for servicing and maintenance. Vertical adjustments control the height in relation to the cutters. This is usually carried out via hand wheels that operate lead screws to raise and lower the table. Stops are fitted to prevent the tables being pushed or wound into the cutter block. The stops should be set so that when the table is pushed back into its working position a gap of no more than 6 mm exists between the cutting circle and table (**Fig. 3.2**).

The front table is adjusted to give different depths of cut (**Fig. 3.3**). While working, it may be necessary to alter the position so as to take more or less cut off the material. Bowed or twisted boards, for example, will need more material taking off than a piece of timber with merely a rough sawn edge.

The back table is adjusted to bring it level with the cutting circle. Failure to set the back table level with (same height as) the cutting circle can present two problems.

Fig. 3.1 Surface planing machine.

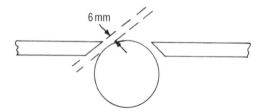

Fig. 3.2 Maximum gap between cutting circle and table.

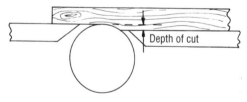

Fig. 3.3 Front table.

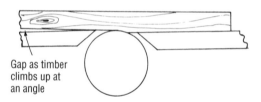

Fig. 3.4 Back table too high.

With the back table too high (**Fig. 3.4**) the material will first of all stop as it hits the table. This is usually enough to alert the machinist to the fact that resetting is required. When the table is only slightly higher than the cutting circle however, the timber will ride over it without any clues to the operator that something is wrong. The first sign that a problem is present is upon completion of the cut when it is found that the end of the timber has not been planed at all. In some cases, when perhaps the operator is inexperienced, a second and sometimes even a third pass over the blades is carried out before cursing the machine, insisting it is broken. No matter how many times the piece is fed over the blades the end will never be planed. This is because, at the end of a piece, both hands will be on the back table pressing it down and causing the last part to lift off the front table

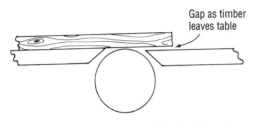

Gap as timber leaves table

Fig. 3.5 Result when weight is placed on back table.

(**Fig. 3.5**). To rectify the fault, switch off and isolate the machine and wait for it to stop completely. With the block now stationary, place a straight edge on the back table and roll the cutter block backwards. As the cutters reach their highest point the bed is lowered until light contact is made with the straight edge. Note that a fine touch is required here to prevent the table being set too far the other way.

With the table too low (**Fig. 3.6**) the work will simply fall or drop on to it as it leaves the front table. This will not only result in a dip on the end of the material (**Fig. 3.7**) but also increase the risk of kick back. Basically, as the timber drops an increase in the amount of cut will result and so increase the resistance to feeding. The lower the table, the more the risk.

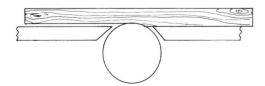

Fig. 3.6 Back table too low.

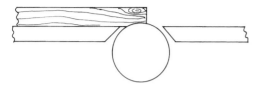

Fig. 3.7 Resulting drop or dip.

Sometimes it may appear that the tables are not set correctly in height when in fact the tables are out of alignment. This is usually as a result of badly worn or poorly adjusted slides which allow the tables to tilt back, thereby preventing them from being parallel to each other (**Fig. 3.8**).

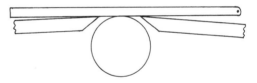

Fig. 3.8 Tables out of alignment.

Most of the older machines could suffer from this problem, but with a little care it does not have to prevent accurate working. First, avoid excessive pressure or leverage on the ends of the tables (particularly when people lean or sit on them). This can cause the tables to tilt even on new machines. When it is identified that the tables are out of parallel they must be reset before any work can commence. In serious cases this will require the table being packed from underneath with spacing shims to bring it level again. In many cases, however, the tables can be reset be winding them down as low as they will go. This levels them as they bottom on the slide. Once they are at the bottom (and level again) simply wind them back into their working positions and check for alignment.

3.1.5 Setting the tables for cutting

For checking the alignment of the tables a good straight edge is essential. A steel rule will be adequate providing that it is straight and without any damage to the edges.

The easiest way is to place the rule on to the back table and carefully roll the block backwards. Be careful not to cut the rule with the cutters. This will not only damage the rule but will also chip the cutters. As the blade meets the rule light contact should be made. The rule should not lift off the table (this would indicate that the table is too low). A faint scraping noise may be heard but it is the slight scraping feeling transmitted through the rule that indicates correct setting. Holding the rule gently between finger and thumb with another finger lightly resting on top of the rule will enhance the sensitivity and allow more accurate setting (**Fig. 3.9**).

Most machines are manufactured with bed inserts (**Fig. 3.10**). Their sole purpose is to allow the edges of the tables to be replaced should they become damaged. The type of damage that usually appears on the inserts is wear as a result

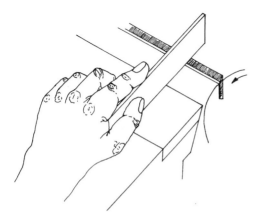

Fig. 3.9 Correct way to check for cutter position.

of continual feeding of abrasive materials, chips as a result of foreign bodies in the material, and dents caused by the spanner used to secure the cutters. If the faults become a problem when working it may be necessary to have the inserts changed, but usually a gentle dressing down with a mill file to remove any lumps or burrs will be sufficient.

3.1.6 Cutter blocks

Placed between the tables is a cylindrical cutter block carrying the cutters or blades. Traditionally there are three main types of cutter block.

a) Cap held blocks (Fig. 3.11)

The cutters are fitted to the block while the cap is off. Fine adjusting screws (usually two per cutter) are set into the block on to which the cutters are fitted via slots (**Fig. 3.12**). Only cutters with slots should be used on this type of block. When the cap is refitted the cutters can be easily adjusted by turning the fine adjustment screws. The cap is held in place by a series of large nuts which gen-

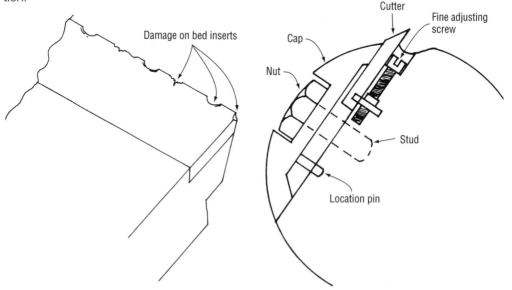

Fig. 3.10 Table inserts.

Fig. 3.11 Cap held block.

erate enough pressure to grip the cutters in place as well.

b) Bar held blocks

These blocks are available in two types. Both work by the same principle, where a steel bar presses the cutter tightly into place. The pressure can be applied by either:

- Allen studs threaded through the body of the block and forcing the bar on to the cutter (**Fig. 3.13**); or
- bolts threaded into the bar which will nip the cutter into place as they are wound out from the bar. As the bolt is wound out of the bar it tightens up against the back of the block, thereby gripping the cutters (**Fig. 3.14**).

Adjusting screws are usually found on both types, but instead of being slotted into the cutters, the cutters simply sit on to a small shoe that carries the cutters up during adjustments. When cutters are being brought higher or out of the block, the shoe will raise them without error. However, if the cutter is too high and needs to go back into the block, these shoes are of little use. As the adjusting block is wound down the cutter will not be taken with it as a slotted cutter would be. Instead, the cutter needs to be tapped on to the top of the adjusting shoe. This can be done by using a wooden tool handle (e.g. screw driver or hammer) and gently tapping the cutters until they sit down on to the adjuster.

c) Wedge bar held blocks

These are very similar to the bar held block, but the bar is now wedge shaped. This type of block is probably the most commonly available largely because it is safer (**Fig. 3.15**). Both the cap held and bar held can have the cutters slip out as the block revolves at high speed but, by making the bar wedge-shaped, as the cutters creep forward the wedge goes with it

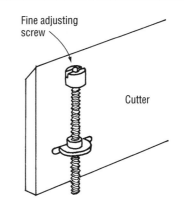

Fig. 3.12 *Fine adjusting screws.*

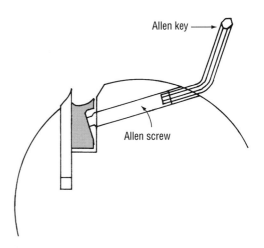

Fig. 3.13 *Bar held block with Allen screws.*

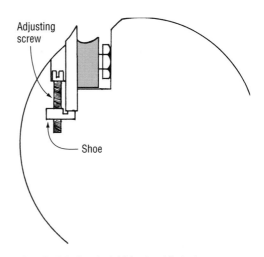

Fig. 3.14 *Bar held block with bolts.*

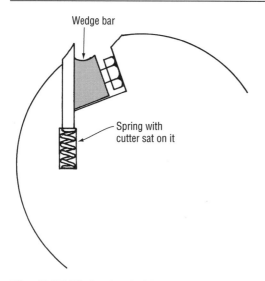

Fig. 3.15 Wedge bar held.

and self-tightens. Adjusting screws, if fitted, are usually the same as those found on bar held blocks.

3.1.7 Fence

The fence is made of heavy gauge steel and fixed to the infeed table (**Fig. 3.16**). It can be wound across the table or cutter by hand wheel through a rack and pinion. This movement is to allow different widths of component to be catered for or simply to find a new, sharper section of blade to feed over.

For facing and edging, the fence must

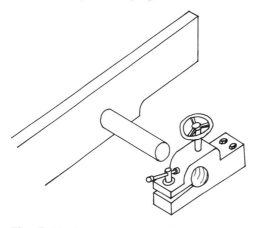

Fig. 3.16 Fence arrangement

be square to the table and this will need checking on a regular basis to ensure accuracy. To check the fence for square, place a metal engineers square on to the bed and up against the fence. Any adjustments can be made after undoing the fence lock. Some machines are fitted with a scale of angles ranging from 90° to 45°, which will make setting easier if it is accurately calibrated.

Even a small amount of out of square will cause many problems later in the manufacturing processes. A little patience is sometimes required to ensure the fence is exactly 90°, but it is essential for successful results on the finished product.

When an angle cut is required (see Section 3.2.6, page 51) a sliding bevel should be used to set the fence in the same way as for setting for 90°. Mark out the required angle on to the work-piece or a board and then set the sliding bevel to the marking out (**Fig. 3.17**).

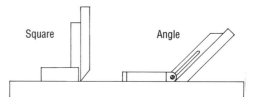

Fig. 3.17 Setting the fence square and setting up for angles.

3.1.8 Guards

All surface planers must be supplied and fitted with two basic guards. These should always be set in accordance with current regulations so as to reduce any hazard associated with the machine.

On the working side a bridge guard is used. The requirements of this guard are as follows.

- It is strong and rigid and made of a suitable material to prevent the ejection of moving or loose parts.
- It is easily adjustable in height and width so that it can be set to accommodate all material types and sizes.

- It is running in parallel and in line with the cutter block.
- It is at least equal in length to the length of the block and cutters.
- It is at least equal in width to the diameter of the cutting circle.

With the above standards met the guard can be set. When setting the guard it must be positioned close enough to the material so as to prevent the operator from making contact with the blades. At the same time a smooth passage should be maintained without the guard causing an obstruction. The following standards are the recommended positions for setting the bridge guard.

3.1.9 Facing and edging

When smoothing or planing the face and edge of material, one operation immediately following the other, the guard must be set with a gap not bigger than 10 mm between the guard and material (**Fig. 3.18**).

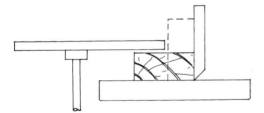

Fig. 3.18 Set-up for facing and edging.

If the material is only going to be planed on the wide surface the guard can be pushed over until it is touching the fence (**Fig. 3.19**). Likewise, if only the narrow edge is to be planed the guard can be lowered until it is resting on the bed (**Fig. 3.20**). By setting this way the cutter is concealed far better, and the risk of accidents is lessened.

When the material is of squared section (the same size in width and thickness) and facing and edging is to be carried out the guard can be set in either of the positions in **Fig. 3.21**. It should

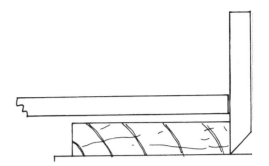

Fig. 3.19 Guard set for facing only.

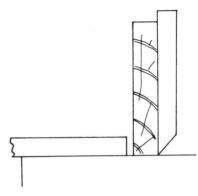

Fig. 3.20 Guard set for edging only.

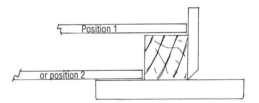

Fig. 3.21 Correct settings for square-section material.

not be set as in **Fig. 3.18** as this would leave a large open area around the block.

Which method you use is entirely down to personal preference, but bear in mind the following points before making your choice:

- With the guard set over the top of the work piece and up to the fence, a large portion of the blade is exposed. Some people see this as a hazard not only to the operator but also to passers by in the event of a trip or fall. This is, however, a common method in the industry because the work can be fed over different parts of the blade to reduce wear.

- The only part of the cutter to be exposed is the part which is cutting the material and if it is cutting material it is difficult to cut fingers at the same time! The only point to remember is that fingers are not allowed in the cutting area (on top of the material, directly above the cutters). See Section 3.2.2 for feeding techniques.

The other guard is known as the block guard or remote side guard. It must be strong and rigid, and be easily adjustable so that it can effectively guard the exposed block behind the fence. The term 'adjustable' refers to the ability to move in and out with the fence as different sized boards are set up for. Ideally it should be fixed to the fence so that it will always be in position without the operator having to adjust it individually.

3.2 Using a surface planer

3.2.1 Setting up for planing

The following procedure is recommended:

1. Set the back table level with the cutting circle.
2. Adjust the infeed table until the desired depth of cut is achieved (**Fig. 3.22**). If a scale is fitted to the machine this can be used to set the position. On machines with no scale use a steel

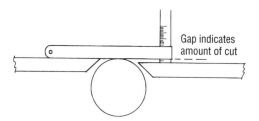

Fig. 3.22 Setting depth of cut.

rule laid on the back table and extending to the front table.

A 1–2 mm gap will be sufficient for most jobs. Wide boards should never have more than this, otherwise the risk of kick back is increased.

3. Set guards around material.
4. Proceed to feed the material over the blades.

3.2.2 Feeding techniques

No matter how well set the blades and tables are, if the material is not fed correctly a poor finish will result.

a) Facing (first pass)

With the machine set and ready to run, look down the material for the best side to cut. Knots and foreign bodies must be watched for but the major concern here is bowed, cupped or distorted board (**Fig. 3.23**).

Fig. 3.23 Sighting up the timber.

b) Bowed boards

Having established which side is bowed, place the hollow side down on to the bed (**Fig. 3.24**). This will allow a stable position and prevent the board from see-sawing as feeding pressure is altered. With boards that are severely bowed it may be difficult to feed them under the bridge

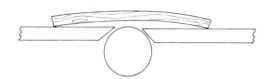

Fig. 3.24 Cupped boards.

guard. In such cases either discard the piece and cut it back for smaller jobs or place the bridge guard flat to the bed and 10 mm (maximum) away from the side of the material. When doing this you cannot plane the face and edge simultaneously unless the material is square section, and also the board cannot be fed at an angle. Feeding at an angle is usual practice on wide boards to prevent the full width of board coming into contact with the cutter in one go. This could cause heavy kick back.

c) Cupped boards

When the boards are cupped, again the hollow side should be facing down during feeding. The timber now has two points making contact with the bed which prevents unwanted movement during feeding.

d) Twist and distortion

This is by far the worst of the three. Even experienced machinists can have difficulty producing a good flat surface when planing twisted boards. The secret is to find where the timber naturally rests when placed on the machine. Then place your hand on it without rocking or disturbing its position. If it does move, simply reposition your hands until you find the right spot. Once found, feed forward until the timber emerges from under the bridge guard on the back table. Remove your left hand and replace it on the timber on the back table. Because the cutters have just planed the emerging timber, some of the twist will have been removed. As soon as you have a firm hold of the timber on the back table you can remove your right hand and place that also on the outfeed end.

If the surface is still unplaned in places the timber will sit on the bed without any problems for a second pass.

3.2.3 Holding the material

The timber needs to be held sufficiently to prevent chattering or vibration as the cutters make contact. At the same time too much pressure must also be avoided, otherwise feeding becomes almost impossible due to the shear weight on top. The best way to describe the amount of pressure is to say 'let the weight of your hands be the pressure'. No exertion of downward force is required under normal circumstances.

The hands must be spread over the board to distribute an even pressure. When watching people work on a surface planer it never ceases to amaze me how many ways people have adopted as feeding techniques. Methods to stay clear of are as follows.

- *The piano player* – who stands facing the infeed table with both hands straight out on the timber (**Fig. 3.25**).

  It is only when half of the timber is under the guard that you realize just how awkward this can be, but by then you are too far involved to do anything about it. It usually results in the operator trying to walk the timber to the outfeed bed after stretching past the bridge guard.

- *Running through* – some people seem to think that both hands must be holding the timber at all times, and as a consequence they leave their hands on top of the material as it passes over the

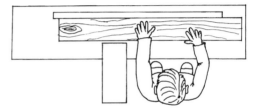

Fig. 3.25 Piano player feeding method.

cutters (**Fig. 3.26**). Even though the bridge guard is set correctly the hands must never enter the cutting area. In the event of kick back or a loss of concentration who knows what could happen?

- *Holding the ends* – when the material is being fed, two fingers (sometimes only one) are placed on the end of the timber and used to assist in feeding by pushing (**Fig. 3.27**). Sometimes a machinist may argue that the timber is sticking, but this is usually as a result of too much pressure. If this is not so and the timber really is sticking, the beds can be lightly rubbed with wax

just as a joiner would do on a hand saw if it were sticking while cutting.

Others may argue that they move their hands from the end before feeding the last bit. Again this may be so, but a lot of accidents have been recorded where the hand was not removed for that last part and the results were planed finger tips.

The following shows a step by step procedure for safe and efficient feeding.

- *Step 1.* Apply light pressure from both hands as the board is entered to the cutters. Feet should be turned to face forward with the body weight evenly distributed (**Fig. 3.28**).

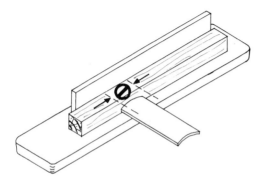

Fig. 3.26 No entry!!

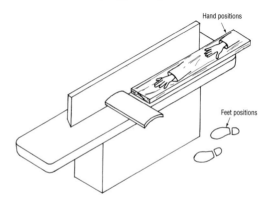

Fig. 3.28 Step 1 of feeding.

- *Step 2.* As the left hand approaches the guard raise the fingers to pass over the top but still leave pressure from the palm or base of the hand (**Fig. 3.29**).

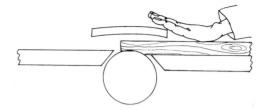

Fig. 3.29 Step 2 of feeding.

- *Step 3.* As soon as the end of the board appears from under the guard transfer left hand to outfeed end and place finger tips firmly down on it. While the

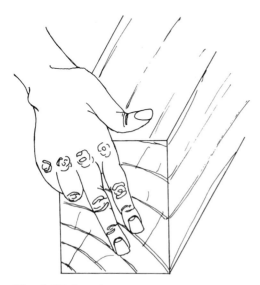

Fig. 3.27 Poor feeding method.

left hand is not touching the timber (during passing over to outfeed end) feeding must continue with the right hand (**Fig. 3.30**).

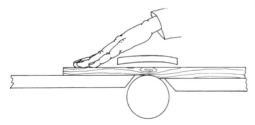

Fig. 3.30 Step 3 of feeding.

- *Step 4.* With both hands flat on the timber, carry on feeding until the right hand meets the bridge guard.
- *Step 5.* Pass the right hand over to the outfeed end and keep feeding with the left hand.
- *Step 6.* Feed hand over hand until the board is fully through.

At all times keep the material moving at a steady, controlled speed. Stopping causes beat marks on the planed surface which are simply created by the cutters rubbing on the timber while it is stationary. Feeding too fast causes ripples on the planed surface (**Fig. 3.31**). As the cutters revolve and make contact with the material a small curve of timber is removed.

Fig. 3.31 Cutter marks on a planing machine.

The faster the timber is fed, the larger the curve will be and this creates a poor finish of ripples or cutter marks on the component (**Fig. 3.32**).

3.2.4 Rebating

A rebate is simply a square or rectangular cut on the corner of a piece of timber (**Fig. 3.33**). The surface planer can cut them very easily and the set-up process is

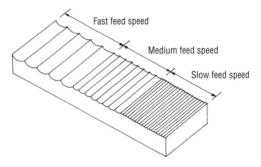

Fig. 3.32 Effect of feed speed on surface finish.

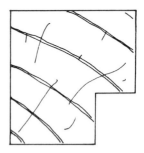

Fig. 3.33 Rebated section.

simple. Rebating is only allowed if the machine is effectively guarded. This requires the use of a tunnel guard which will provide a pressurized tunnel around the cutting area through which the component is fed. The conventional bridge guard is removed to allow the tunnel guard to be set in position around the material (**Fig. 3.34**).

A stopped rebate (see shaping machines page 113, **Fig. 5.45**) cannot be cut on this machine. It would need the top pressure of the tunnel guard to be removed to allow dropping on and this would prevent the cutting area from being effectively guarded.

The tunnel guard is set by first positioning the side pressure. Set it to an approximate position by using adjuster 'A' so that the pressure pad is close to the material and then lock off the nut. Then undo the nut on adjuster 'B' and firmly press the pad against the material. It must be tight enough to secure the work-

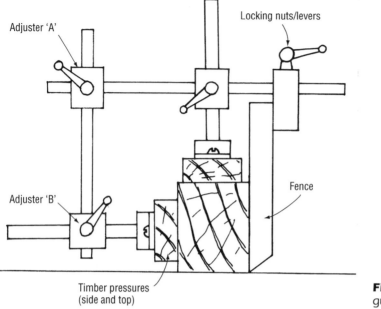

Fig. 3.34 Typical tunnel guard.

piece up to the fence but not so tight that feeding is difficult.

When the top pad is set, again ensure enough pressure to hold the material down to the bed but not too much. There are now two pads pressing on the material, and if feeding is difficult the operation becomes dangerous. Overexerting causes uncontrolled feeding and could result in jerking the feed which creates a hammering effect on the cutters.

Hammering on to the cutters is also more likely when rebating because a deeper cut is usually being made. A recommended depth of cut is 10 mm. If a job requires much more than this, the material should be removed in two or more passes. Also, the depth of cut should be adjusted in relation to the width of cut. For example, a wide rebate being cut on a door frame (44 mm × 12 mm) would be best made in two passes (**Fig. 3.35**) whereas a small glass rebate (24 mm × 12 mm) could safely be made in one pass (**Fig. 3.36**).

In order for the machine to cut a rebate the cutters must be set to stick out beyond the edge of the out feed table.

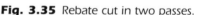

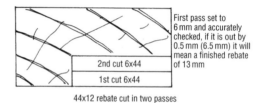

First pass set to 6 mm and accurately checked, if it is out by 0.5 mm (6.5 mm) it will mean a finished rebate of 13 mm

2nd cut 6x44
1st cut 6x44

44x12 rebate cut in two passes

Fig. 3.35 Rebate cut in two passes.

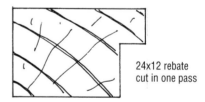

24x12 rebate cut in one pass

Fig. 3.36 Rebate cut in one pass.

Failure to do this would mean that the timber will catch on the out feed bed and slowly be pushed away from the cutters, resulting in a tapered rebate.

The cutters must also be ground to allow a clearance from the cut surface. The corner of a new cutter would usually be supplied with no side clearance (**Fig. 3.37**). If it is intended that the machine is going to be cutting rebates it must be

Fig. 3.37 New cutter.

ground as shown in **Fig. 3.38**. This will allow a very sharp point to cut the side of the rebate, resulting in a clean, smooth cut. Without this clearance the grain will break away from the corner as the cutter leaves the material, and burn marks will be left on the cut surface.

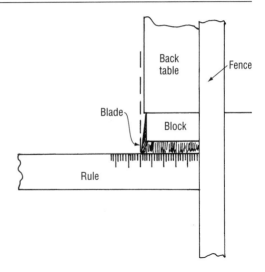

Fig. 3.39 Setting the fence for width of rebate.

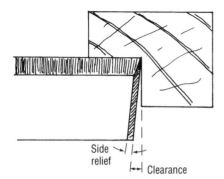

Fig. 3.38 Correctly ground for rebating.

3.2.5 Setting up for rebating

The following procedure should be adopted.

1. Check and adjust the back table as necessary. It must be level with the cutting circle.
2. Move the fence so that the distance from the edge of the blade to the fence is equal to the widest dimension of the rebate (**Fig. 3.39**). To assist in setting, a steel rule can be placed against the fence and then simply wind the fence in or out until the cutter point is on the correct measurement on the rule. Sometimes this operation is easier to do by replacing the steel rule with a square of plastic. The rule may be too narrow to butt up to the fence and still reach to the cutter. Mark the size of

the rebate on to the plastic with a pencil and wind the fence accordingly.
3. Lower the front table to suit the required depth. If the scale on the machine is accurate simply read off it and proceed with next stage. Where the scale is unreliable use two steel rules. Lay one across the back table reaching over the block to the front table. The second rule is then stood up on the front table and where the two rules intersect gives the depth of cut.
4. Set the tunnel guard around the material and ensure that free and easy movement exists.
5. Make a test cut. Make any necessary adjustments, after isolating the machine.

3.2.6 Chamfers and bevels

Just like cutting rebates, any kind of bevel or chamfer can only be cut if a tunnel guard is used. The compulsory use of tunnel guards for cutting rebates, chamfers and bevels is relatively new. Traditionally, they could be cut without any safety devices other than the bridge guard. As a consequence many 'old hands' or experienced machinists still work without the tunnel guard. This is in fact illegal and serious repercussions can result from breaking regulations.

3.2.7 Setting up for cutting bevels and chamfers

This should be carried out as follows.

1. Check and adjust back table as necessary.
2. Set a sliding bevel up to the required angle. This can be set from the setting out rod or directly off the marked out component (**Fig. 3.40**). Ensure that the blade of the bevel does not protrude beyond the handle or stock. This would allow inaccuracies when sitting it on the bed for fence setting (**Fig. 3.41**).

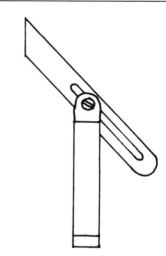

Fig. 3.41 Wrongly set bevel.

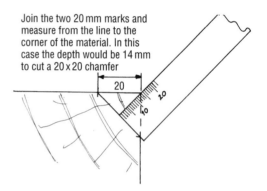

Join the two 20 mm marks and measure from the line to the corner of the material. In this case the depth would be 14 mm to cut a 20 x 20 chamfer

Fig. 3.42 Measuring angles for depth of cut.

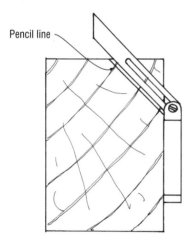

Pencil line

Fig. 3.40 Setting of sliding bevel.

3. Undo the fence lock and adjust the fence angle. The sliding bevel should be placed with the stock on the front table and the blade butted up to the fence. When satisfactory, lock the fence in position and recheck. On many machines the process of locking the fence causes it to move slightly.
4. Set the front table to the depth of cut. This is not as simple as it is for rebating. The exact depth has to be found by measuring the material as shown in (**Fig. 3.42**).

5. Set the tunnel guard around the material.
6. Make a test cut and check angle.

On all rebate, bevel and angle cutting, if the front bed or fence have to be adjusted to change the size or angle of a cut, the tunnel guard will not need subsequent adjustments. This is thanks to the fact that it is fitted to the fence which is in turn fitted to the front table. As either is moved the tunnel guard moves with it.

A very common mistake made when rebating or angle cutting is trying to make a second cut when the first cut has not cut quite deep enough. After passing the material over the cutters, for example, the bevel is found to be 1 mm too small (**Fig. 3.43**). The operator now drops the front table an extra 1 mm to correct the fault and makes a second cut (**Fig. 3.44**).

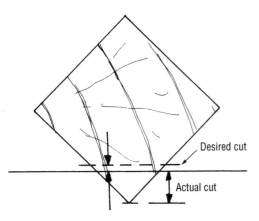

Fig. 3.43 First cut too small.

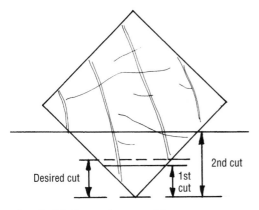

Fig. 3.44 Second cut too big.

With this explained you are probably thinking how obvious this is and how stupid people must be who do this and you are probably correct. I have made this mistake and every year I see several others do it!

3.2.8 Sheet materials

It is possible to prepare the edge of sheet materials on a surface planer, but always use the same part of the cutter and in a place along the cutter that will not be required for everyday use.

Sheet materials contain very abrasive materials such as glue, which blunts the cutters almost instantly. When planing plywood, for example, a series of lines are left on the cutter where the glue lines touched (**Fig. 3.45**).

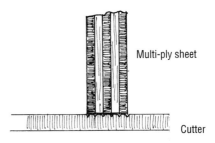

Fig. 3.45 Chipped cutters after planing multi-ply.

To prevent this damage spoiling the finish on all following pieces of timber, keep all sheet materials to the far side of the cutters.

Alternatively, tungsten carbide tipped cutters can be bought which will stand up to the harsh abrasives found in sheet materials and some hardwoods. They are more expensive to buy, but on such materials they will far outlast conventional high speed steel (HSS) cutters and are therefore worth the extra initial outlay.

It is not recommended that TCT cutters are used in situations where softwood is the most widely used material. TCT is hard and will withstand abrasive materials, but it is also brittle. To keep the points from breaking off, the cutter is made less sharp by reducing the grinding angle. (This is similar to altering hook and sharpness angles on saw teeth.) For softwoods this reduced sharpness would produce a surface finish known as woolly. Basically, the resinous grain has lifted like thousands of hairs on the surface. HSS cutters are very sharp (30–35° angle) and cleanly slice through the timber to remove the tiny hairs.

Some timbers, such as teak and mahogany, are renowned for having interlocked grain. This is where, during the growth of a tree, the fibres that make up the timber have twisted around each other or crossed paths and become intertwined. When machining with conventional HSS cutters the grain can break or be plucked, leaving a broken surface where the grain has snapped away. With the introduction of TCT cutters the reduced grinding angle helps to reduce the breaking of the grain.

Modern cutter blocks work in a shearing motion where a thin strip of tungsten carbide is inserted around the block. This gives a progressive cut which reduces breakout even further. (This is covered in more detail when discussing jointing machines, page 83, Tapered seating blocks.)

3.2.9 Safety devices

To assist in the feeding of awkward and irregularly shaped components push blocks can be used (**Fig. 3.46**).

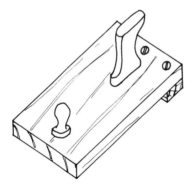

Fig. 3.46 Push blocks.

The base-board must be of a similar width and length as the component being planed. A wooden cleat is fixed to the base which the work piece is butted up to during feeding. The cleat can either be glued and screwed from above or set into a groove and glued. Metal parts should not be allowed to make contact with the cutters and neither should the cleat. Always ensure that the material is thicker than

the height of the cleat from the baseboard. In the event of the cleat touching the cutters severe kick back will result.

A good example of using a push block for feeding irregularly shaped items is when smoothing a round board, such as a coffee table top (**Fig. 3.47**).

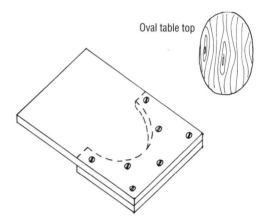

Fig. 3.47 Push block for irregularly shaped components.

3.2.10 Cutter setting

There are several commonly adopted methods for setting cutters accurately in a surface planer. They all produce the same end result, which is to set the cutters parallel to the back table.

a) Precision setting device

Usually supplied with the machine, this device is probably the most accurate of all the methods. In fact, to set the cutters properly can take a considerable amount of time and is sometimes considered too accurate (**Fig. 3.48**).

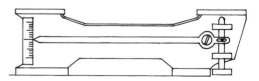

Fig. 3.48 Precision setting device.

The procedure for use is as follows.

1. Ensure the bed is zeroed on the scale.
2. Position the setting device on the left-hand side of the back table with the spring-loaded pad directly above the top of the cutter block.
3. Place a mark on the front foot of the setting device where it intersects with the edge of the table. This is to ensure it is positioned in the same place when repositioned.
4. With firm finger pressure holding the device, carefully roll the cutter block backwards. The spring-loaded pad will be pushed up, which will cause the pointer or needle to move along the scale. Never roll the block forward as this will allow the cutters to bite into the metal pad and chip the cutters.
5. As the cutter reaches its maximum height check where the needle registers on the scale.
6. If the needle does not read zero, make adjustments by using the cutter adjusting screws to raise or lower the blades accordingly. It may take a few adjustments to get it exactly on zero.
7. Move the device to the opposite side of the table (right-hand side) and ensure that the pencil mark on the foot is lined up with the edge of the table.
8. Repeat steps 4, 5 and 6 until both sides of the cutter are registering the needle at zero.

The reason why this method is considered too accurate is that if one side is only fractionally lower than the other, the needle will exaggerate the amount (**Fig. 3.49**).

In most cases a tiny difference is acceptable but when viewed on the scale it looks pretty drastic.

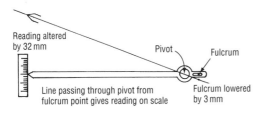

Fig. 3.49 Exaggerated accuracy.

b) Hardwood setting strip

This is a very simple method that gives good, quick and accurate results. Before starting with this method it is worth running the hardwood strip over the surface planer to provide a new, straight edge to work with. Recommended procedure for use is as follows.

1. Lower the back table by 1–2 mm. Some people keep it level but this will prevent the cutter from getting a hold on the strip.
2. Lay the strip on the back table (again to the left-hand side) and allow it to extend beyond the cutter block. Make sure that the body of the block cannot touch the strip or a false reading will result.
3. Place a pencil mark on the strip where it intersects with the back table. Feel free to number or letter the marks as you apply them so as not to become confused when several marks are present (**Fig. 3.50**).
4. Roll the block forward by hand. As the cutter touches the strip it will lift it up and carry it forward. Make a second line where the strip is laid down (**Fig. 3.51**).
5. Move to the opposite side of the table and repeat steps 3 and 4 with the original line (numbered 1) being on the edge of the table to start with. If line 2

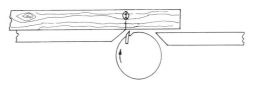

Fig. 3.50 Setting the strip.

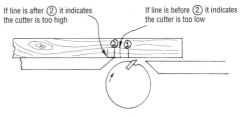

Fig. 3.51 Second position.

does not stop on the edge of the table the cutter is not parallel.

6. Adjust cutter height as necessary and recheck until both ends are the same. When they are the same, reset the table to zero on its scale.

c) Steel rule

This method is by far the quickest and easiest, and if done correctly it is accurate enough. In industry it is probably the most commonly used method. The procedure for use is as follows.

1. Ensure the back table is set at zero on the scale.
2. Place a steel rule on the back table so that it overhangs the cutter block.
3. Roll the block by hand backwards. Listen and feel on the rule for contact with the cutters. If the rule is lifted off the bed the cutters are too high. If no contact is observed then the cutters are too low. How sensitive the finger tips are, will determine the accuracy of this method.
4. Repeat step 3 on the opposite end of the cutter and adjust as necessary.

Note that heavy contact with the cutters could chip or blunt them.

3.2.11 Removing and replacing cutters

The recommended procedure is given below.

1. Isolate the machine and remove the fence and guards to allow easy access.
2. Undo the table locks and pull the tables clear from the block.
3. Position the block so that the cutter locking nuts can be reached and then wedge the block in place. A pair of wooden wedges is sufficient, one at each end of the block. This will prevent the block slipping as pressure is exerted to slacken the nuts.
4. Undo the locking nuts fully or until the cutters can be removed. On a cap-

held block the nuts need to be removed completely to enable the cap to be lifted off the cutters. On bar-held blocks the nuts only need slackening to allow the cutters out.

5. Remove the cutters and the holding devices. Remove any build-up of resin from all components. The cutters need only be clean enough to prevent the resin clogging the grinding wheel when being reground. Usual practice is to place the other components in a tub or bath of solvent to soften the build-up. Also clean the block and cutter seatings. Dust and resin often build up here and failure to remove them will prevent accurate setting later.

6. Using a clean rag, wipe all surfaces to remove any surplus of solvent on the block or components.

7. Check all parts for damage or wear and replace as appropriate.

8. Refit the cutters to the block and set into approximate position.

9. Reset the table to its correct position.

10. Accurately set the cutters using your preferred method. The cutter locking nuts will only be finger tight at this stage, to allow adjustments to take place.

11. When the cutters are accurately set, tighten the nuts fully in the correct sequence. Ideally you should start in the centre and work out, alternating from left to right until all the nuts are tight. Other acceptable methods are to start at the end (left or right) and work across to the other side (**Fig. 3.52**). By working in one of these three methods any twist bow or distortion in the cutters is flattened out. If a bow is trapped in the middle after the two ends have been tightened, the steel has nowhere to go.

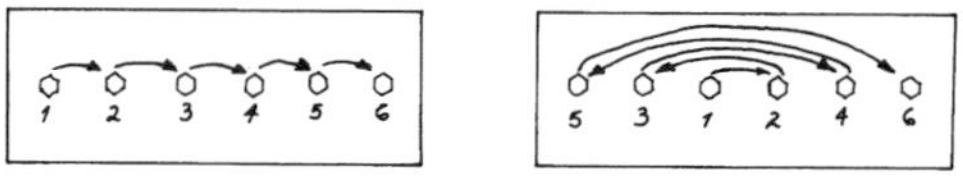

Fig. 3.52 *Acceptable tightening methods.*

As the centre nuts are tightened the stress could cause the cutter to slide out of position, thereby undoing the setting just done (worse, it could crack the cutter).

3.3 Thickness planer

3.3.1 Basic functions

The basic function is to plane timber, which has been surfaced and edged, to a finished width and thickness.

3.3.2 Secondary functions

With the aid of jigs, bed pieces or saddles, tapering and bevelling can be carried out successfully.

3.3.3 Design and layout

The solid machine casting carries the cutter block, feeding rollers to drive the material through, pressure to hold the material and a height-adjustable bed or table (**Fig. 3.53**).

3.3.4 Bed

The bed is adjustable via a hand wheel, or on modern machines an automatic switch. Either way an accurate scale is essential for quick and easy referencing when working. The most commonly used scales consist of a rule fixed on to the machine casting, and a pointing device fixed to the machine bed. As the bed is raised or lowered the operator reads off the position of the pointer on the scale until the desired size is achieved (**Fig. 3.54**).

When using this type of scale a test cut or measuring of the first piece is essential to ensure accuracy. The same operator could end up with slightly different reading through simply standing in a different place when setting. It is best to bend down and read the scale at eye level to reduce errors. More modern machines incorporate digital and even LCD read outs and switches to control the movement. These prove to be much more reliable and consistent (**Fig. 3.55**).

The bed should also be fitted with antifriction rollers to aid feeding by preventing the material sticking to the bed. Few machines will work efficiently without them but some manufacturers do not fit them as standard equipment. By the shear weight or force applied to the material, as the rollers and pressures hold it, feeding becomes difficult. If less pressure were used then it would not be held sufficiently. To overcome the problems free running rollers are strategically placed in the bed. To be of any real value the rollers must be fitted directly below the top rollers (**Fig. 3.56**).

As the material is entered into the machine the top infeed roller will grip it and begin to feed it under the block. At the same time, contact is made with the first antifriction roller and the material will be pinched between the two rollers.

For general machining the antifriction rollers are set approximately 0.5 mm above the height of the bed. If the material is wet, resinous, rough sawn or just soft, they may need raising. Initial setting is achieved by placing a steel rule on the bed and feeding it forward until it makes contact with the roller. The gap under the rule indicates the roller height.

Adjustment is made by a hand wheel on the front of the bed and although a scale of measurements is not marked on, calibrations are present to assess how much movement has taken place (**Fig. 3.57**).

Some machines, particularly newer ones, tend to make the adjustments via a lever. The end result is the same where the rollers are raised or lowered but now a scale can be created. Some machines use a numbering system where position 1 would allow the rollers to be below the bed for use with bed pieces. Position 2 would be the standard 0.5 mm above the bed for general work, and with the lever

Fig. 3.53 Typical thickness planer.

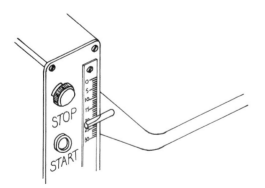

Fig. 3.54 Setting the scale.

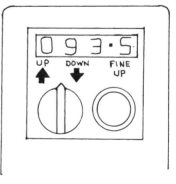

Fig. 3.55 LCD readout and adjusting switch.

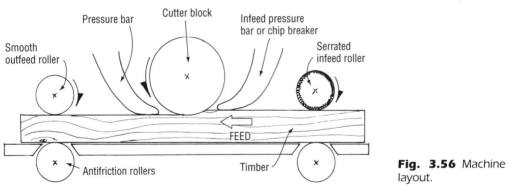

Fig. 3.56 Machine layout.

in position 3 the rollers would be set at 1–2 mm above the bed for feeding problem materials (wet, rough sawn etc.). Bigger heavier-duty machines provide a series of marks that the lever should be lined up with, but no numbers to indicate height. After a few days of running the machine you would become fully aware which positions were best suited to particular types of work.

If the material does not feed easily the rollers can be raised slightly more, until smooth feeding takes place. In fact, some operators set the rollers by first lowering them below the bed and then feeding the material in. They then raise the rollers until the feeding is smooth and uninterrupted.

3.3.5 Feed rollers

Just as for the antifriction rollers, if not set properly, material feeding will present problems. If the rollers are set too high the material will not be held secure and the surface finish will be rippled where chattering has taken place. Too much pressure will either make feeding difficult of simply squash the material.

a) Infeed roller

Two types are commonly available. The solid feed roller and the sectional feed roller. Both are steel serrated rollers but sectional rollers are far superior.

When feeding material on a solid feed roller the whole roller jumps or springs

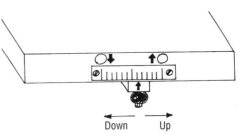

Fig. 3.57 Antifriction roller height adjuster.

up to the height of the material. This presents no problems when feeding one piece at a time, but if more than one piece is entered into the machine at the same time only the biggest will be held securely (**Fig. 3.58**).

The regulations with which we work to do not allow such a machine to feed more than one piece at a time unless an anti-kick back device is fitted. As the cutters spin round at high speed any material not held tightly by the roller would be ejected or kicked out of the machine. This is where the kick back device does its job, by catching the piece and preventing it from being thrown out.

Made from either tough rubber or metal, they are shaped and positioned to allow the material to enter but not to

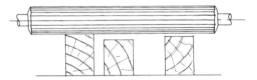

Fig. 3.58 Feeding problem with solid rollers.

be ejected from the infeed end (**Figs 3.59–3.61**).

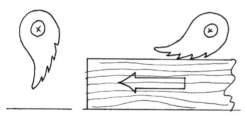

Fig. 3.59 Kick back fingers in rest position.

Fig. 3.60 Pushed forward as timber enters.

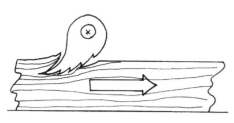

Fig. 3.61 Spikes dig in as timber is kicked back.

The sectional feed roller, however, can easily accommodate several pieces being fed at the same time. This is because the roller is made up of a series of sections or segments all fixed in a line. Each section is independently spring loaded so that it can jump up to suit the individual heights of the pieces of timber (**Fig. 3.62**).

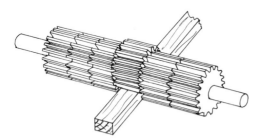

Fig. 3.62 Sectional feed roller.

Although it is not a legal requirement, it is still advisable to have a kick back device fitted with this type of roller. In the unlikely event that two pieces of varying width come under a single roller, kick back would still be present (**Fig. 3.63**).

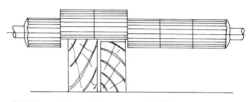

Slightly smaller sections are not held if kept too close together

Fig. 3.63 Problem when pieces are too close.

b) Outfeed roller

These, unlike the infeed roller, are smooth surfaced to reduce the amount of marking left on the cut material. Some manufacturers even rubber coat them to reduce marking even further, yet still give a good grip. Sectional outfeed rollers are not required as the material has passed under the cutters and all pieces will be a standard height.

Both infeed and outfeed rollers should be checked on a regular basis for chipping and for resin deposits building up. If the infeed roller becomes clogged with waste the serrated teeth will no longer bite into the material to feed it through the machine. Also, the individual sections can become fused or stuck together by resin.

The outfeed roller is less of a problem. First, it is not serrated but smooth, so waste cannot build up as easily. Second, the roller is usually fitted with a scraper plate, which if set correctly should remove any build-up before it becomes a problem. Do keep a look out for tell-tale signs of build-up, however, as a small chip of wood can leave several depressions on the finished surface as the roller repeatedly turns.

Brushing with paraffin and oil will soften the deposits, then either scrape or wire brush them off. Remove any surplus oil from the bed and other parts with a rag before feeding any material.

3.3.6 Pressures

Curling around, under the block, the infeed pressure bar serves a second function, that of a chip breaker. By reaching beyond the block it can firmly press down on the material as the cutters chip away to remove unwanted timber. They can be solid or sectional, to match the type of infeed roller, but both are spring loaded to allow a variation of stock thickness.

The very nature of timber produces some weird and wonderful grain patterns that can look very handsome on a finished product. As a circular cutter block spins round to cut away the timber the grain can split or run. As a curl or shaving is lifted the cutter can wrench or rive the grain along its length.

With a pressure bar firmly pressing down on the material as close to the cutters as possible, the riving of the grain is greatly reduced.

The outfeed pressure bar has no effect on the grain but it is very important for holding the material down on to the bed to prevent chattering during the cut. This is especially so when the material leaves the infeed pressure.

Most machines will be set to the factory setting upon delivery to a customer but sometimes a machine may need resetting or adjusting. This could be due to the age of the machine or simply personal preferences in roller and pressure positions. In the furniture industry, for example, the infeed roller is usually set level with the cutting circle, and in the joinery industry the same roller is set 1 mm below the cutting circle. This seems such an insignificant amount that neither industry would notice the difference but both have their reasons.

The joinery industry does not work to such fine tolerances as the furniture industry does, and so if a window or door component is 0.5 mm bigger than the other components it would not matter. On such a large item it would probably not even be noticed. On a decorative chest of drawers, however, all components must be correct, as this is a show piece and many eyes will be cast upon its beauty.

If 0.5 mm were to be removed from a piece of timber set up for joinery use the sharp serrated teeth would bite into the timber deeper than the amount to be removed. The result is a series of marks on the finished surface. Had the rollers been set level with the cutting circle no such marks would be present.

3.3.7 Setting the pressures and rollers

To set the machine (**Fig. 3.64**) for use as preferred by the furniture industry, use the following sequence.

1. Prepare two pieces of timber as deep and as wide as possible (220 mm × 70 mm).
2. Adjust the rollers and pressures so that they are above the cutting circle and the anti-friction rollers are below the bed.
3. Lay the setting pieces on the outsides of the bed and running through the machine. Light pressure would be useful here, which can be achieved by using cramps. This will prevent the pieces moving.
4. Raise the table until the cutters just touch or graze the surface of the setting pieces when the block is turned by hand.
5. Lower the infeed roller and pressure to just sit on the surface of the pieces. Make sure that both sides are the same.
6. Lower the table by 0.25 mm.
7. Lower the outfeed roller and pressure until it just rests on the surface of the setting piece.
8 Lock all adjustments off to prevent movement.

For a machine to be used mainly for joinery work (**Fig. 3.65**) proceed as follows.

1–4. As for furniture industry set-up.
  5. Lower the bed by 0.25 mm.

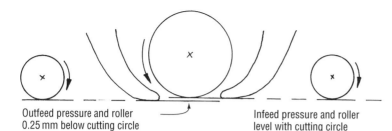

Outfeed pressure and roller
0.25 mm below cutting circle

Infeed pressure and roller
level with cutting circle

Fig. 3.64 *Setting for furniture industry use.*

6. Lower the outfeed roller and pressure until it just rests on the surface of the setting piece.
7. Lower the bed a further 0.75 mm, making a total of 1 mm below the cutting circle.
8. Adjust the infeed roller and pressure so that they just rest on the surface of the setting pieces.

There are many different machines on the market and many different ways of adjusting the rollers and pressures. Therefore always refer to the maker's instruction in the machine manual when making adjustments. A series of locking nuts, wing nuts and springs can be very confusing, but with reference to the manual all will come clear. If the manual is unavailable a process of trial and error must take place. In such an event take a few minutes to analyse the machine and make notes and diagrams relating to:

● what adjustment you need to make;
● which screws or nuts adjust which pressures;
● what adjustment you have made;
● what the original setting were.

Remember, if it needs adjusting in the first place something must be wrong, so you cannot make it any worse.

3.3.8 Drive systems

Traditionally, a single electric motor was responsible for driving the feed rollers and block within the machine. Obviously the block has to travel much faster than the feed rollers, so a gearbox must be incorporated to slow down the speed created by the motor to a more suitable speed for the rollers (**Fig. 3.66**).

The motor runs at a fixed speed that drives the cutter block through vee belts and pulleys. The same motor has a second pulley wheel that drives the gearbox through a special toothed or segmental link belt. The feed speed is altered by moving the gear lever at the front of the machine. Usually, three speeds are available of approximately 6, 12 and 18 metres per minute (position 1 on the lever is the slowest). Some machines also incorporate a stepped pulley wheel on the gear box which will then allow six different speeds (6, 12 and 18 plus 9, 15 and 27 metres per minute).

Adjustments to the feed speed should be carried out while the machine is stopped, to avoid crunching the gears. Moving into neutral is acceptable while the rollers are moving, especially if the feed needs to be halted immediately. Switching the machine off will allow the

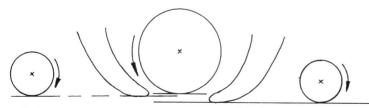

Outfeed pressure and roller
0.25 mm below cutting circle

Infeed pressure and roller
1 mm below cutting circle

Fig. 3.65 *Setting for joinery use.*

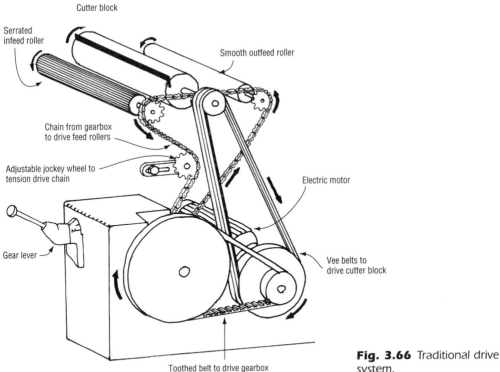

Cutter block

Serrated infeed roller

Smooth outfeed roller

Chain from gearbox to drive feed rollers

Adjustable jockey wheel to tension drive chain

Electric motor

Gear lever

Vee belts to drive cutter block

Toothed belt to drive gearbox

Fig. 3.66 Traditional drive system.

feed rollers and cutter block to run down for several seconds before coming to a stop.

Modern machines have done away with the three-speed gearbox in favour of an infinitely variable drive. In simple terms this is a wide vee belt in between two expanding pulleys (**Fig. 3.67**).

To change the speed, simply turn the hand wheel until the desired speed is displayed. Again, the speed can only be altered while the pulley wheels are in motion.

3.4 Using a thickness planer

3.4.1 Changing the blades

The actual techniques of removing and refitting blades is identical to that on a surface planer but, just as for the surface planer, there are several different setting methods. The method used can be selected by the operator but is usually determined by the manufacturer of the machine.

Integrated setting devices are usually easy and provide an accurately set cutter. Typical examples are given below.

a) Setting rollers

The brass roller, held in a pivoting arm, is swung into position on top of the pre-set adjustment screws (**Fig. 3.68**). When the blades are replaced into the block they should just touch the rollers (one at each side of the machine). It is necessary to hold the pivoting arm firmly in place when checking cutter positions. Failure to do so could result in the arm lifting off the adjustment screw and giving a false reading. Also, be careful not to turn the block and cutters into the rollers. The cutters could chip or become blunt but, more so, repeated blows will damage the rollers rendering them useless when trying to set accurately.

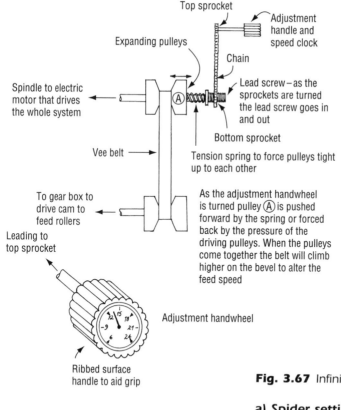

Top sprocket

Adjustment handle and speed clock

Expanding pulleys

Chain

Spindle to electric motor that drives the whole system

Lead screw – as the sprockets are turned the lead screw goes in and out

Bottom sprocket

Vee belt

Tension spring to force pulleys tight up to each other

To gear box to drive cam to feed rollers

As the adjustment handwheel is turned pulley Ⓐ is pushed forward by the spring or forced back by the pressure of the driving pulleys. When the pulleys come together the belt will climb higher on the bevel to alter the feed speed

Leading to top sprocket

Adjustment handwheel

Ribbed surface handle to aid grip

Fig. 3.67 Infinitely variable feed drive.

a) Spider setting device

By far the quickest and easiest, and probably also the most accurate, method is the spider (**Fig. 3.70**). In its most basic type the spider is made up of two steel rods with feet and a pad fixed to each end. As the feet settle on to the body of the block, the cutters are pushed on to the setting pads by springs. In a matter of seconds the cutters can be accurately set in place.

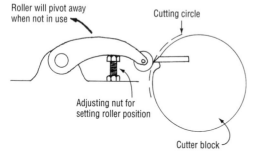

Roller will pivot away when not in use

Cutting circle

Adjusting nut for setting roller position

Cutter block

Fig. 3.68 Setting rollers.

b) Location pin and setting bolt

This is usually a bolt-on device rather than a permanent fixture of the machine. The location pin is pushed into the gap between the wedge bar and block which sets the cutter block into position so that the cutters will touch the setting bolt. The cutters are lightly pushed by springs until they touch the brass bolt heads (**Fig. 3.69**).

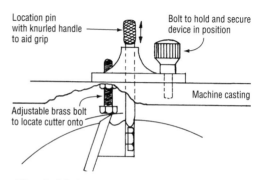

Location pin with knurled handle to aid grip

Bolt to hold and secure device in position

Machine casting

Adjustable brass bolt to locate cutter onto

Fig. 3.69 Location pin setting device.

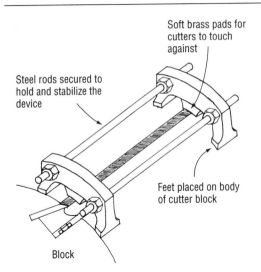

Soft brass pads for cutters to touch against

Steel rods secured to hold and stabilize the device

Feet placed on body of cutter block

Block

Fig. 3.70 *Spider setting device.*

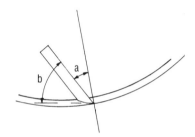

Fig. 3.72 Cutter with back bevel and angles.

3.4.2 Cutting problem timbers

As previously mentioned in this chapter, the grain pattern produced by timber as it grows can have a serious effect on the ease of cutting. For normal grain as is usually found in pine (also known as redwood, deal, Scots pine etc.) a standard cutter (**Fig. 3.71**) will produce good results. With a more dense and abrasive timber but still straight grained, a secondary bevel can be applied to reduce the sharpness of the cutter (**Fig. 3.72**). By making the cutter more blunt it also becomes stronger and is therefore less likely to chip or snip when coming into contact with the abrasive material.

When timber with interlocked grain is being planed a front bevel can be applied. Sapele is a timber recognized by its stripy, interlocked grain. A standard

cutter would tear and rive the grain causing large voids in the surface. By applying a front bevel the fibres of the wood are scraped away rather than being cut.

Any bevel should only be applied when it is identified that the standard cutter cannot achieve an acceptable finish and if there are a large number of components to cut. They can be easily applied by either a slip stone or on a grinding machine, but to remove them requires a full regrind.

3.4.3 Bed pieces and saddles

To utilize the thickness planer further, simple bed pieces and saddles can be used to carry out otherwise impossible tasks. Window beading, subcills and tapered table legs to name a few, are typical examples of work done with the assistance of bed pieces and saddles.

Figure 3.73 is a typical bed piece. The

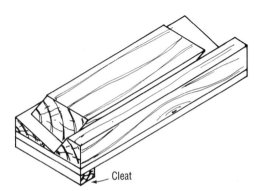

Cleat

Fig. 3.73 Typical bed piece.

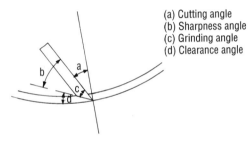

(a) Cutting angle
(b) Sharpness angle
(c) Grinding angle
(d) Clearance angle

Fig. 3.71 Standard cutter.

cleat butts up to the machine bed to prevent it travelling through. The work pieces are fed through the machine by laying them on top of the bed piece. Because the antifriction rollers will not be in contact with the material, sticking is often experienced while feeding. To try to overcome this, operators often glue strips of plastic laminate on to the surface of the bed piece or rub on candle wax.

Unlike a bed piece, the saddle (**Fig. 3.74**) is fed through the machine while it

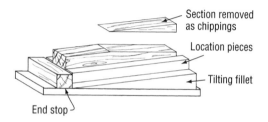

Fig. 3.74 Typical saddle.

carries the component at the required angle.

Self-assessment

The following questions have been written around the previous text in this chapter. If you cannot answer any of the questions, simply restudy the respective areas. Good luck!

1. Which of the following is the correct setting of the back table on a surface planer when setting up for facing and edging?
   (A) slightly lower than the cutting circle
   (B) slightly higher than the cutting circle
   (C) level with the infeed table
   (D) level with the cutting circle.

2. Which is the safest type of cutter block as used on a surface planer?
   (A) cap hold
   (B) bar hold
   (C) wedge bar hold
   (D) square.

3. What is the primary function of a surface planer?
   (A) to take to width and thickness
   (B) to plane two adjacent sides square
   (C) rebating
   (D) chamfering.

4. What purpose does a chip breaker serve on a thickness planer?
   (A) to prevent large pieces breaking out along the grain
   (B) to apply pressure to hold the material down
   (C) to prevent the cutters being ejected
   (D) both (A) and (B) combined.

5. Which of the following should be set the lowest when the thickness planer is set up for joinery work?
   (A) cutter block
   (B) infeed roller
   (C) outfeed roller
   (D) outfeed chip breaker.

6. What are antifriction rollers used for on a thickness planer?
   (A) feeding dry timber
   (B) to prevent timber sticking while feeding
   (C) to prevent timber splinters wedging into the bed
   (D) to help feeding when using jigs.

7. Upon inspecting the first piece after setting up the surface planer it is found that a small depression or dip is found at the end of the timber as it leaves the machine. Explain the most likely cause of this fault.

8. Explain the sequence of tightening the securing nuts on the block of a thickness planer.

9. Explain when it may be necessary to apply a front bevel to thickness planer cutters.

10. Explain why a surface planer must never have a front bevel applied to the cutters.

Crossword puzzle
Planing machines

From the clues below fill in the crossword puzzle. The answers are all related to the surface planer or thickness planer.

Across

1. Standard safety device to prevent fingers touching the blades.
4. Used to remove the timber, slice by slice.
7. Angle that is put on to the end of a cutter when it will be cutting a rebate.
9. Height at which the cutters must be set at in relation to the back table.
10. Used to take away the waste produced when cutting.
11. Safest type of planing block.
12. Right angle cut on the edge of timber.
16. Slip stone the cutters.
17. Creepy crawler for setting cutters.
18. Opposite of 18 down.
19. Length, width and ---------.

Down

1. Jig or saddle used to hold or carry the material through a thicknesser.
2. Prevents machine from starting up.
3. The antifriction rollers will be doing this when the material passes over them.
5. Safety device which must be used when cutting rebates.
6. The rules by which all guards are set.
8. Can be angled but is usually at 90° to the bed and cutters.
13. Carries the cutters.
14. Flat or baseball-type of holding device on a cutter block.
15. Raise or lower the infeed bed to vary this.
18. When finished the machine is turned ---.

4 Jointing machines

4.1 Hollow chisel morticer

4.1.1 Functions

Mortice machines are used to cut a slot, or mortice hole, into a piece of timber to house the tenon (**Fig. 4.1**). The mortice hole can be:

- through – where it passes all the way through the material;
- stub – where it is stopped inside the material;
- haunched – a combination of the above to standard proportions;
- angled – either through, stub or haunched.

4.1.2 Design and layout

Mortice machines can vary immensely from one model to another, even when made by the same manufacturer. The basis of all machines is a casting or frame which carries a bed and fence to locate the work onto. A centrally mounted electric motor drives the cutting medium while it is brought into the work piece by a hand lever on the right-hand side. A series of hand wheels (usually three) are used to move and position the work piece (**Fig. 4.2**).

4.1.3 Bed

The bed has three different adjustments, all worked by hand wheels.

a) Longitudinal

Longitudinal or sideways movement is used to traverse the bed left and right (**Fig. 4.3**). The amount of movement is assessed by the operator who simply stops upon reaching the marking out lines or by the use of adjustable table stops. Stops are particularly useful for production runs when it would not be cost effective to mark out each individual piece. There are many different types of stop available. For the type and use of stops on your machine refer to the manufacturer's instruction booklet.

The mechanics of this sideways movement is very simple. A steel strip with teeth (known as a rack) is bolted to the

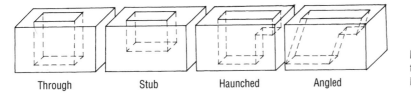

Through   Stub   Haunched   Angled

Fig. 4.1 Basic types of mortice hole.

Fig. 4.2 Hollow chisel morticer.

Again a very simple principle allows this movement. It is basically a nut and bolt. The bolt is connected to the bed with a hand wheel on the end of it and the nut is set into the machine casting. As the hand wheel is turned the bolt will travel through the nut and carry the bed with it. A locking screw is sometimes situated to the side of the hand wheel to prevent the position of the fence being altered by turning the wrong hand wheel.

c) Rise and fall

This facility, which is not available on all machines, is used to set the depth of cut.

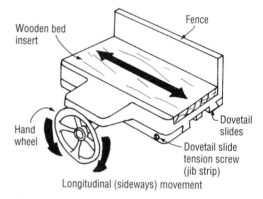

Fig. 4.3 Longitudinal (sideways) movement.

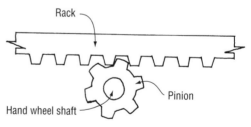

Fig 4.4 Rack and pinion.

underside of the bed and a cog (or pinion) is fixed to the end of the hand wheel shaft. As the hand wheel is turned the teeth on both parts interlock and travel over each other (**Fig. 4.4**).

b) Lateral

Lateral movement (in and out) is used to position the fence and bed in relation to the chisel, to suit the marking out (**Fig. 4.5**).

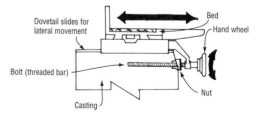

Fig. 4.5 Lateral movement (in and out).

The movement is worked through splayed gears or, less often, helical gears (**Fig. 4.6**).

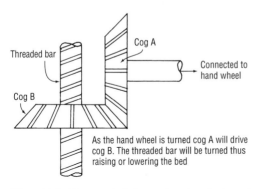

As the hand wheel is turned cog A will drive cog B. The threaded bar will be turned thus raising or lowering the bed

Fig. 4.6 *Rise and fall mechanics (splayed gears).*

4.1.4 Headstock

The headstock carries an electric motor that drives the tooling. It is mounted on to the casting via dovetail slides to allow smooth easy movement and, through adjustment of tension screws, any play in the movement is eliminated (**Fig. 4.7**).

The headstock is brought into the work piece by pulling on the hand lever situated to the right of the machine. The movement in the hand lever is transferred to the headstock by one of two methods.

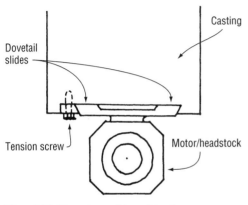

Fig. 4.7 *Plan view of headstock.*

a) Counterbalance

Usually found on older machines, this design relies on weights being hung from a wire rope and connected to the top of the headstock via pulleys (**Fig. 4.8**). The weight should be equal to the headstock so that it will rest in any position. This is particularly useful when setting up depth stops but it will also make working for long periods easier as there will be little or no resistance when moving the head in either direction.

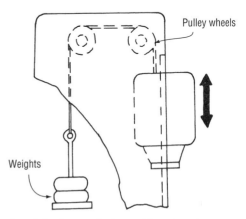

Fig. 4.8 *Counterbalanced headstock.*

b) Rack and quadrant

Similar to the sideways movement of the bed, a quadrant of teeth engage the teeth of the rack as the hand lever is pulled. This in turn pushes the head down the slide (**Fig. 4.9**). To even out the weight of the head and stop it from falling unexpectedly, a spring counterbalance is connected to the head. The tension of the spring can be adjusted by turning a ratchet set into the casting.

Where large amounts of headstock travel are necessary the hand lever can be repositioned to suit the operator. This is done by loosening the locking nut on the end of the hand lever and adjusting as necessary on the splines (**Fig. 4.10**).

4.1.5 Depth stops

Most machines, if not all, will have some means of controlling the depth to which

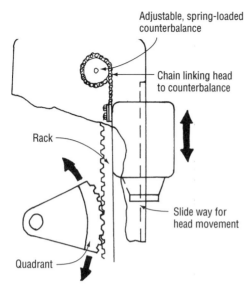

Fig. 4.9 Rack and quadrant.

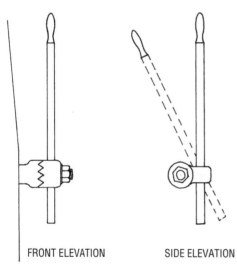

FRONT ELEVATION   SIDE ELEVATION

Fig. 4.10 Hand lever.

the head travels into the material. On some machines there may even be provision for more than one setting. This is needed when more than one depth is to be cut into the material, such as a mortice and haunch.

a) Single depth stop

When the tooling is set into the machine the head is brought down until the bot-tom (points) of the chisel is level with the lines marked on to the timber. The stop bolt is then raised until it reaches the stop block and then tightened. Note that if the bolt is overtightened it will distort the stop bar and restrict headstock movement.

As the head is lowered the contact switch is released, allowing the motor to start. On return, the switch is pressed back in and the power is cut off thus stopping the motor (**Fig. 4.11**).

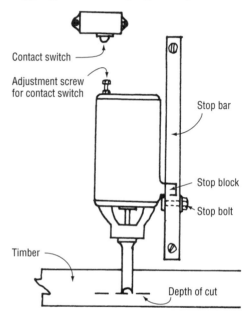

Fig. 4.11 Single depth stop.

b) Multiple depth stops

This type of stop is set in the same way as the single stop by moving the chisel points to the marking lines but, instead of a bolt, steel discs or collars are set to touch the underside of the stop pin (**Fig. 4.12**). The collars are locked into place with Allen screws which pass through the collars to nip on to the stop bar.

When the collars are set for the two depths required they can be selected by moving the stop pin in or out. Collar B is selected for the shallow holes by pushing the pin in, and collar C for the deeper holes by pulling the pin out.

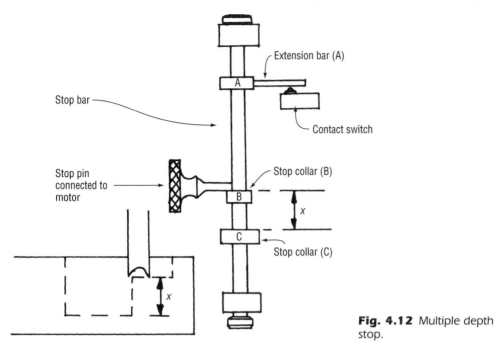

Fig. 4.12 Multiple depth stop.

The machine will automatically start when the head is lowered and the extension bar A presses the contact switch. When the head is returned the switch is released and the power is turned off.

Fig. 4.13 Hollow chisel.

Fig. 4.14 Auger.

4.1.6 Tooling

In most cases there are four components required for this type of morticer.

a) Chisel

This should be as close as possible to the size of hole being cut and is measured from point to point across the face. It must have at least one slot in its side to allow waste to escape (**Fig. 4.13**). This slot is known as the chip ejection window.

b) Auger

This runs through the centre of the chisel and should be paired up with it when bought so as to ensure a perfect fit (**Fig. 4.14**). It should be free from bends along its length and the spiral should extend far enough to reach the chip ejec-

tion window. Positioned at the bottom of the auger are two ears or wings which score the surface of the timber. Also at the bottom are two flats, one to each side of the centre. Following the scoring of the surface, the flats will scoop out the waste.

c) Bush

This is used to centralize and hold the chisel firmly in place. Two types are commonly used to suit the make of machine. Split bushes (**Fig. 4.15**) rely on a pinch bolt, passing through the machine casting, to close the split and allow it to grip the chisel shank. Plain bushes (**Fig. 4.16**) have a hole in them which allows an

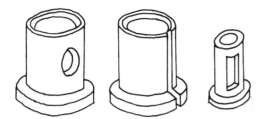

Fig. 4.15 Plain bush.   **Fig. 4.17** Collet.
Fig. 4.16 Split bush.

Allen screw to pass through and grip the chisel shank directly.

d) Collet

This is used to centralize and hold the auger. It works exactly like a plain bush, with the screw gripping the stem of the auger (**Fig. 4.17**).

If you try to imagine a 12 mm mortice chisel opened out it would be equal to a 48 mm joiners chisel which would be extremely difficult to drive through timber. For this reason the chisel does no more than clean and shape the hole. To permit this the cutting circle of the auger wings must be slightly bigger than the distance across the face of the chisel, thereby allowing the auger to remove the bulk of the waste (**Fig. 4.18**).

As the auger spins, the waste produced is drawn up its spiral and released through the chip ejection window. To allow the waste to enter the spiral a clearance gap must be left when setting up (**Fig. 4.19**). For chisels up to 12 mm (0.5 in) the wings of the auger should be approximately 1 mm lower than the points of the chisel. Chisels over 12 mm

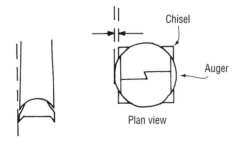

Fig. 4.18 Relationship between chisel and auger sizes.

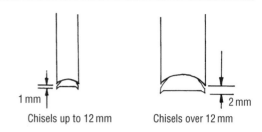

Fig. 4.19 Clearance gaps.

should have approximately 1.5–2 mm clearance. When the timber is resinous more clearance will be required.

Should the wrong combination of the four parts be selected or the correct combination be set wrongly, several problems can arise.

- *Collet too big for auger.* As the Allen screw is tightened there will be no support from the collet until the auger is forced back into it. The result is that the auger will be bent, causing it to spin out of parallel to the chisel and rub on it. This will create friction as the two metal surfaces rub together which will lead to heat being generated and the chisel turning blue around the centre.
- *Bush too big for chisel.* This chisel will be set out of parallel to the auger, giving the same results as above.
- *Auger too small for chisel.* The auger will not be guided by the chisel because of the large gap inside. The chisel will be working very hard to remove the remaining waste. The results are that feeding the tooling into the timber will be very difficult, due to the chisel having to remove so much waste. The chisel will blunt quickly and the auger could distort or bend.
- *Auger spiral too short.* Chippings and waste will not be expelled through the ejection window. As a result, waste will build up inside the chisel and eventually clog up. This will generate heat as the auger revolves and could cause the chisel to turn blue at the centre. Continued use could cause the chisel to split.

- *Auger too high in the chisel.* The two metal surfaces will rub together and generate heat. This can be identified at an early stage by listening for a squeaking noise as the auger revolves. The bottom of the chisel will blue and crack. These cracks can run up the chisel and cause one or more of the points to break off.
- *Auger too low in chisel.* This will allow the auger to cut larger chips of waste which could clog the inside of the chisel, particularly if the timber is resinous. This results in clogging of the waste inside the chisel, leading to overheating. Another problem with this set-up is that the auger will cut much deeper than the chisel and make depths of cut inaccurate (**Fig. 4.20**).

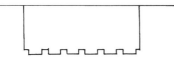

Fig. 4.20 Result of auger being set too low.

4.2 Using a hollow chisel morticer

4.2.1 Fitting the chisel and auger

The following procedure should be adopted.

1. Select the correct combination of components to suit the work being done. Check that all parts are in good working order.
2. Assemble the four components together as shown in **Fig. 4.21**. Ensure:

   (a) the flat on the auger lines up with the hole in the collet and both are in line with the Allen screw on the casting;
   (b) the chip ejection window is facing

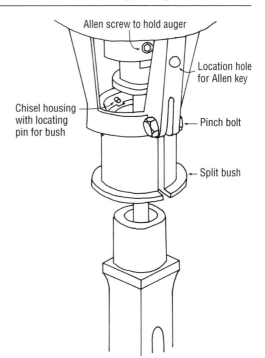

Fig. 4.21 Tooling assembly.

to the right of the machine, to allow the waste to be quickly expelled away from the direction of cut (left to right);

   (c) the hole in the chisel bush lines up with the Allen screw for tightening in place – when a split bush is being used, a guide pin is used to ensure that the split line up with the pinch bolt.

3. With the machine isolated, insert all components into the machine. Allow the auger to drop by approximately 10 mm and nip the Allen screw on to the auger (**Fig. 4.22**). With this accomplished none of the other parts can fall because the hole through the chisel is smaller than the end of the auger.
4. Push the chisel and chisel bush up into the casting and square the chisel off the fence by using a small metal try square (**Fig. 4.23**). If the chisel is not set square to the fence the finished

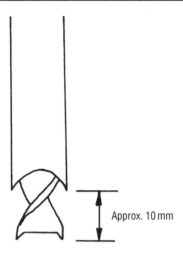

Fig. 4.22 *Auger dropped by 10mm.*

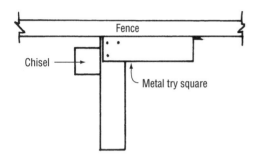

Fig. 4.23 *Squaring the chisel.*

mortice hole will be 'stepped' (**Fig. 4.24**).

5. Tighten the chisel into place with the Allen screw or pinch bolt.
6. Loosen off the auger and raise it into the chisel, remembering to allow the required clearance.
7. Spin the auger by hand and listen for any squeaking or scratching. These are sure signs that the chisel and auger are rubbing.

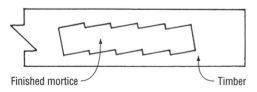

Fig. 4.24 *Stepped mortice (chisel not square to the fence).*

8. Now that the tooling is set a test cut can be carried out.

4.2.2 Setting up for a stub mortice (to marking out)

Even when using stops, at least one piece of timber needs marking out to give the initial settings of the machine. With this marked out piece set the machine in the following order.

1. Position the work piece on the machine bed with the face against the fence and set the cramp to hold it firmly. (Note that too much cramp pressure could compress the timber while being cut. When released, the timber will resume its original shape and you may find the mortice hole bigger than the chisel used to cut it.)
2. Adjust the bed height so that the timber is about 30mm below the chisel when in the rest position. This will prevent excessive travel of the headstock. The headstock hand lever may also need adjusting to be within comfortable reach.
3. Bring the chisel to within 1mm of the top of the timber and turn the in and out hand wheel so that the back of the chisel is level with the marked line on the timber (**Fig. 4.25**).
4. Traverse the bed so that the chisel clears the timber when brought down to the depth of cut (**Fig. 4.26**).
5. Bring chisel down to the pencil line on the end of the timber and set the stop

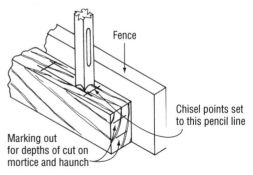

Fig. 4.25 *Setting chisel from the fence.*

Fig. 4.26 Chisel free to be lowered.

bolt or collar to prevent the head travelling below this point (**Fig. 4.27**). If the spring counterbalance is correctly tensioned the head will stay in any position without you having to hold it. This allows you to use both hands for setting the stops.

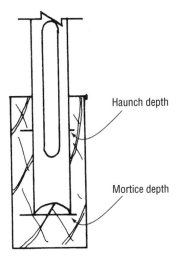

Haunch depth

Mortice depth

Fig. 4.27 Chisel lowered to pencil lines.

6. Cut a test piece and check the position from the face side and the depth of cut.

4.2.3 Cutting a through mortice

This type of mortice is commonplace on joinery items such as windows and doors. There are two ways of carrying out a through cut.

a) Two cuts (one from each side)

1. When the machine and tooling are set proceed to cut the mortice hole to

approximately two-thirds of the way through (**Fig. 4.28**).

2. Now turn the piece over so that the underside is facing up. The piece must be turned end-to-end to keep the face up against the fence.

3. Cut into the piece until no timber is left inside the mortice hole (**Fig. 4.29**).

When cutting through in this way marking out lines need taking on to the underside and great care must be taken to cut exactly to them. Failure to cut to the lines will result in steps inside the mortice hole (**Fig. 4.30**)

b) Single cut

Simply set the machine and tooling in the usual way, but set the depth stop to allow the chisel to pass through the timber completely. It is necessary to use a false

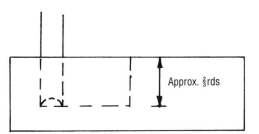

Approx. $\frac{2}{3}$rds

Fig. 4.28 First cut of a through mortice.

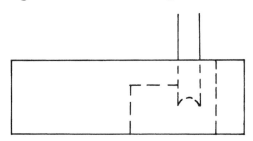

Fig. 4.29 Final cut of a through mortice.

Step

Fig. 4.30 Step caused by poor marking out or careless cutting.

bed when through cutting, to reduce the breakout on the underside of the mortice as the chisel comes through. Also use bed stops to keep each piece in the same place. This will also reduce break-out by supporting the timber at the point where the chisel breaks through. Note that if the timber is put in different positions the chisel will cut into the false bed and reduce the support to the timber.

This method is particularly useful on mass production runs as the work will be completed much quicker. The surface finish on the underside, where the chisel comes through, is poor in comparison with the two-cut method. For this reason it is advisable only to carry out this type of morticing on windows and door frames where the poor finish will be tight up against brickwork and therefore unseen.

4.2.4 Angled mortices

Angled mortices can be carried out quite easily with the aid of a saddle or bed piece. Simply cut out the angle required from a piece of timber or sheet material and sit this on to the machine bed. Mount the timber on to this saddle and cramp into place (**Fig. 4.31**).

There are two points to consider.

1. Excessively steep angles are impossible on most machines, because either the cramp will not be able to hold them or the timber will not fit on the machine due to it fouling on some part of the headstock.
2. Depth stops cannot be used for the

majority of angled work because they work in parallel to the bed. As you move further up the slope you will be cutting deeper into the material.

To overcome this problem mark the front of the chisel with a fine point, permanent marker pen or scratch the mark on with a needle file (**Fig. 4.32**).

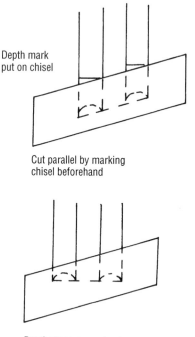

Depth mark
put on chisel

Cut parallel by marking
chisel beforehand

Depth stops prevent cut
from being parallel to
edge of timber

Fig. 4.32 *Depths of cut on angled mortices.*

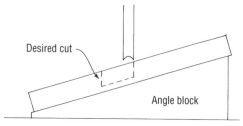

Fig. 4.31 *Cutting angled mortices.*

Desired cut

Angle block

4.3 Single-ended tenoner

4.3.1 Basic functions

As its name implies, the single-ended tenoner is used to cut a tenon to suit a precut mortice hole. This can be quite simply a tenon with a straight, square shoulder (**Fig. 4.33**) or, on the opposite side of the scales, angled shoulders with double scribes (**Fig. 4.34**).

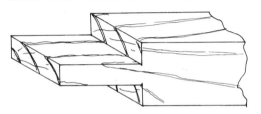

Fig. 4.33 Basic tenon.

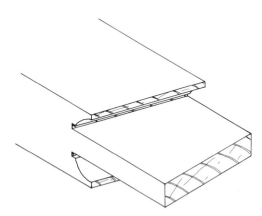

Fig. 4.34 Angled shoulder, double-scribed tenon.

4.3.2 Design and layout

The most common and most practical machine consists of a top and bottom tenon head, followed by top and bottom scribing heads and finally a cut-off saw (**Fig. 4.35**).

Larger machines have the heads mounted on to a swan neck casting. This allows the heads to be evenly distributed and, more importantly, allows long tenons to pass between the motors without fouling on the casting (**Fig. 4.36**).

When space is limited or the nature of the work being carried out does not require this clearance between the motors, a smaller machine is available where the heads are mounted on to a pillar casting (**Fig. 4.37**).

4.3.3 Machine table

Onto the main framework or casting, a hardened steel track is accurately positioned which allows a sliding table to feed the work through the cutting heads (**Fig. 4.38**).

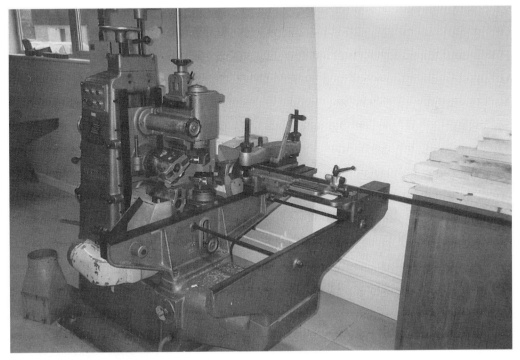

Fig. 4.35 Tenoning machine.

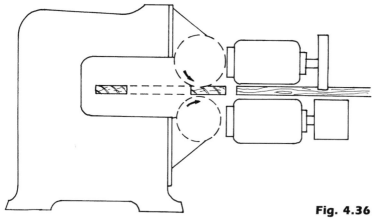

Fig. 4.36 Swan neck casting.

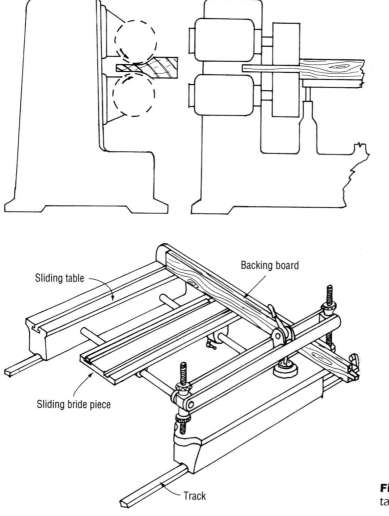

Fig. 4.37 Pillar casting.

Backing board

Sliding table

Sliding bride piece

Track

Fig. 4.38 Sliding table arrangement.

This sliding table has an adjustable steel fence bolted to it which will, from time to time, need checking for accuracy of 90° to the cutting heads. (To check this setting simply cut a tenon and hold a try square against the shoulder and face edge. If it proves to be out of square, slacken off the locking nut and swing accordingly from the pivot point.)

It is essential that a wooden backing piece is fixed to the fence as this will prevent break-out of the timber fibres as the cutting heads chop through (**Fig. 4.39**). Ideally this backing piece should extend far enough into the machine to prevent break-out or spelching along the whole tenon, but in practice most machinists set it just past the shoulder, where most of the break-out occurs. Any break out along the tenon will be hidden inside the mortice hole and will consequently be of little concern.

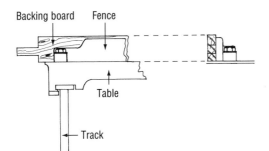

Fig. 4.39 Steel fence and backing board.

4.3.4 Component stops

Usually, a machinist will only mark out one piece of timber or component because, once a machine is set and locked in position, any number of components can be fed through and be identical to the ones before it. Provision is also made on tenoning machines to reproduce identical component or shoulder lengths. There are three main types fitted to the machine during its manufacture:

a) Dead stop

This is used to control the length of tenon on the first cut end of the component. Once it is set to the marking out lines, all subsequent pieces are simply butted up to it (**Fig. 4.40**).

b) Shoulder stop

This is a spring-loaded steel strip set into a sliding bridge piece (**Fig. 4.41**). Once the first end is cut using the dead stop, the tenon shoulder is butted up to the strip. (The tenon shoulders are usually more accurate to work from as opposed to the end of a tenon.) To make adjustment to the shoulder length a locking lever is situated under the bridge piece which, when slackened off, will allow it to slide between the table rails. When not in use the spring-loaded strip can be

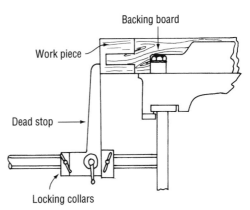

Fig. 4.40 Dead stop.

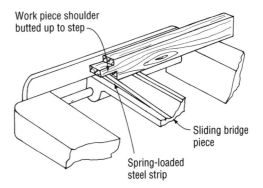

Fig. 4.41 Shoulder stop.

retracted into the bridge piece by tightening the wing nut below it.

c) End stop

The end stop is positioned on the fence via a stop bar (**Fig. 4.42**). It is an alternative to the shoulder stop and is used for controlling component lengths that extend beyond the bridge piece. They are worked similarly to the shoulder stop but can be set to the shoulder or the end of the work piece.

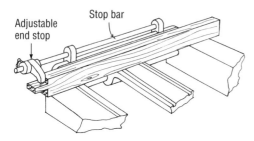

Fig. 4.42 End stop.

When the length of the component reaches beyond the normal stops homemade stops can be set up. A typical type is to fix a length of timber to the fence with G cramps and then cramp a block to the length of timber to act as a stop. This approach can be awkward, and sometimes the timber can flex and give an inaccurate reading. However, with a little care successful results can be achieved.

4.3.5 Cramping arrangements

Cramps vary according to the manufacturer of the machine and the customer's specifications. The two shown here are the most common. Both can be either hand operated or modified to work with the assistance of compressed air (known as pneumatic). When operating as manual units, both systems work with the aid of eccentric cams (**Fig. 4.43**). When converted to pneumatic systems, an air cylinder or piston is used to force the pads on to the work piece (**Fig. 4.44**).

A problem regularly experienced is cramping short components that do not allow full cramping pressure. A quick solution is to use a cramping strip, which is basically a scrap piece of timber used to lay on top of two components (**Fig. 4.45**). When using dry, smooth hardwoods I strongly recommend that coarse sandpaper is glued on to the cramping strip to give a better grip. Once this has been done the same strip will do for all similar cramping jobs.

4.3.6 Cutting heads – tenoning blocks

For cutting tenons there are two main types of cutter block. These are the tapered seating cutter block (also known as shear cut or helical curved blocks) and the rectangular block. On both types the heads are fitted to the machine with one

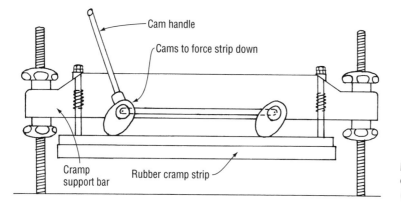

Fig. 4.43 Multiple component cramp (eccentric cam).

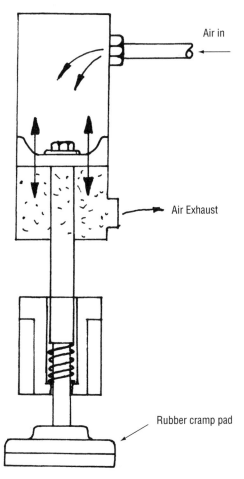

Fig. 4.44 *Pneumatic, single component (all carried on a cramp support bar).*

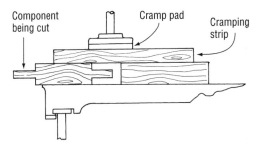

Fig. 4.45 *Cramping strip for short components.*

head in front of the other so as to reduce the cutting impact as the cutters make contact with the timber. If both heads were to strike the timber at the same time

the timber could chatter or vibrate as it is being cut and result in a poor finish.

a) Tapered seating blocks

The first point to realize with this type of cutter block is that the cutters which produce a flat tenon are actually curved! This is necessary because the faces of the block where the cutters are seated are not parallel to each other. They are, as the name implies, tapered. This meaning that the distance between the cutter seatings at the shoulder end of the block will be smaller than at the bearing end.

Before this illogical cutting principle is understood, consider the following similar, yet simple examples.

When using a smoothing plane to shoot the edge of a piece of timber the plane may be held at an angle to the timber, but still fed in a straight line. This will produce what is known as a shearing or paring cut, giving a much easier removal of waste (**Fig. 4.46**).

Similarly, a joiner shears or pares away waste with a chisel by feeding it at an angle. Again the job is made much easier and a better finish will result (**Fig. 4.47**).

Fig. 4.46 *Shear cutting with a hand plane.*

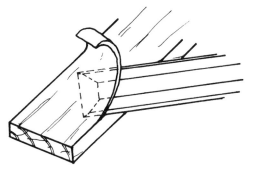

Fig. 4.47 *Paring cut with a chisel.*

Another example of shear cutting is found when examining a cylinder lawn mower. As the blades rotate only a very small portion will actually be touching the grass. This small amount of grass is therefore cut with great ease and, as the other portions of the blade progress to make contact, a full width of cut is made. Usually lawn mower blades are not sharp but will cut easily due to this progressive cut and, more importantly, because the blades are shearing or paring the grass as with the hand tools (**Fig. 4.48**).

A tenon cutter works along the same principles, by progressively shearing its way along the tenon face. To allow this shearing action the cutters must resemble the blades of a lawn mower. However, because we cannot bend the cutters around the block, a slice is removed from it and the cutters are ground to the required shape (**Figs 4.49** and **4.50**).

Fig. 4.48 Cylinder lawn mower.

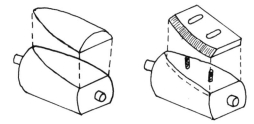

Fig. 4.49 Cutter block with slice removed. **Fig. 4.50** Cutter with helical curve.

It is important to note that cutters with helical curves should not be swapped from one block to another. If blocks have different dimensions (length, diameter etc.) the helical curve will also be different. For this reason every tapered seat cutter block is supplied with a steel grinding template to match the cutter to when grinding/resharpening (**Fig. 4.51**). In the event of the template being lost or damaged the cutter shape can be geometrically developed as shown in **Fig. 4.51**. In the following description, reading this text by itself may confuse rather than help. It is a good idea to plot out the instructions as they are read, even if only rough sketches.

1. Draw the block and cutting circle as required. (both must be sat on the same centre line).
2. Project a line upwards from the shoulder end of the block.
3. Take a line from the top of the cutting circle until it meets the shoulder line at point A.
4. From point A draw a line parallel to the block face to reach the bearing end at point B.
5. Split the block face line into equally spaced divisions. (Eight are used in **Fig. 4.51** but any number can be used. The more lines you use the more accurate the finished drawing will be but it can get very confusing if too many are used.)
6. Project lines 1 to 8 up on to line A–B and also across to meet the cutting circle.
7. Using a pair of compasses, swing the lines round to meet the shoulder end line.
8. Square the lines on the cutting circle up and the lines on the shoulder end across until both meet.
9. Join the intersecting points to form the required helical curve.

b) Rectangular blocks

This type of block was developed during the late 1950s (**Fig. 4.52**). It was thought

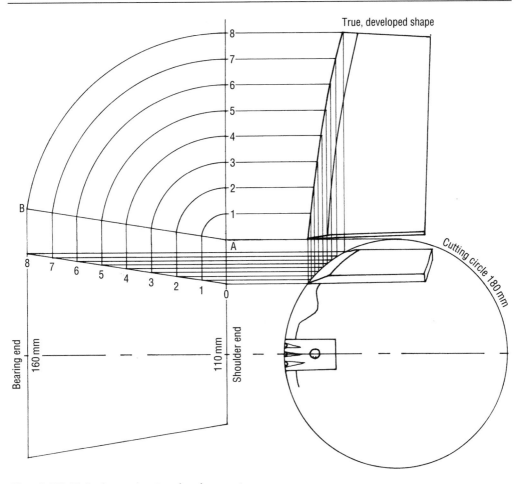

Fig. 4.51 Helical curve cutter development.

that a flat blade producing a constant angle of approach into the timber would produce a much cleaner cut with less vibration instead of an ever changing approach angle as used by a shear cut block. Experimentation established that a 50° cutting angle was required but, so that the point would not be weak and break off, the shape of the block had to be changed from square to rectangular. By doing this the cutter produces a 50° cutting angle but can still be ground with a standard grinding angle (about 30° depending on the type of timber) to prevent the point becoming weak.

In practice there is little to separate the finished product, whichever block is used. Both produce a good finish on what has always been considered a problem area.

One problem associated with both blocks is the size of shavings produced. When cutting long tenons the shaving comes away from the material as a thin flake equal to the length of the tenon which could block up the extraction pipe. On larger machines the blocks are split, comprising four cutters on two independent units (similar to trenching heads, see cross-cut section page 14). On smaller machines where the blocks are supplied in one piece the operator can eliminate the problem by inserting a gap or chip along the cutting edge, which will

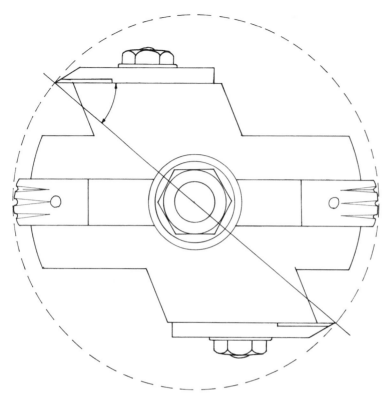

Fig. 4.52 Rectangular cutter block.

leave a bead of uncut material along the tenon. The shaving is therefore broken as it comes away from the material.

4.3.7 Spur cutters

When cutting tenons we are cutting across the grain and this can produce a very poor finish, in particular along the shoulder where the grain breaks out or spelches as the cutters leave the timber. In order to prevent this break-out, spur cutters (also known as shoulder cutters or lansing irons) are fitted to the blocks to sever the fibres cleanly. They must be set in advance of the main tenon cutters, both in width and depth (**Figs 4.53** and **4.54**).

Note that the performance of a spur cutter relies firstly on it being positioned correctly on the block, but also that the shape and sharpness are maintained.

Templates should be used to keep the desired shape during grinding. Also note that a spur cutter should only be ground on the inside edge (**Fig. 4.55**).

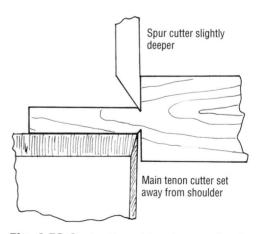

Spur cutter slightly deeper

Main tenon cutter set away from shoulder

Fig. 4.53 Spur cutter set in advance of main tenon cutters.

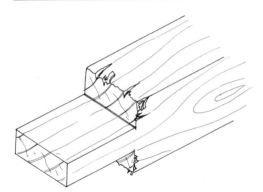

Fig. 4.54 Broken grain (Spelching) caused by tenon cutter too close to shoulder or spur cutter not deep enough.

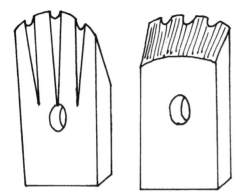

Fig. 4.55 Typical spur cutters.

4.3.8 Changing tenon and spur cutters

The following steps are the correct procedure for changing these cutters.

1. Prepare a piece of hardwood PAR (planed all round) approximately 1 m × 195 mm × 70 mm. This will be used as the setting template.
2. Set the machine tenoning heads to cut a tenon on to both the top and bottom of the template. The timber needs to be positioned to allow the whole length of the cutter to be registering on it, but there is only a need to cut a depth of between 5 and 10 mm.
3. Cramp the material firmly and cut the tenon on to the template.
4. Switch off and isolate the machine,

keeping the template firmly cramped in place.

5. Remove the top and bottom guarding arrangements to allow easy access to the tenon heads.
6. Remove the top tenon cutters and spur cutters and thoroughly clean the block. This may require the application of a cleaning agent or solvent to soften the built-up resin before it can be easily scraped off. In place of expensive cleaning fluids a 50:50 mix of paraffin and oil can be used. It is important, however, that neat paraffin is never used on machine parts because it contains water. By mixing with oil a protective coating will be applied to the machine, thereby preventing corrosion.
7. Wipe away any residue of paraffin and oil or solvent to allow a good contact between the components.
8. Place the spur cutters and securing bolts into the positions provided on the block and tighten the bolt finger tight.
9. With both spurs on the block, slide one of them down to touch the setting template lightly. The points should sit in the scribed line made earlier and a light resistance should be felt when rolling the block. To ensure accuracy of the position of spur cutters the following test can be carried out.

   - Release the setting template and reposition it so that the spur cutters are in line with the centre of the tenon face.
   - Turn the block by hand to allow the spur to score a mark or groove across the tenon (**Fig. 4.56**).
   - Mark the entry and exit of the spur and measure the distance between. Ideally this should measure 16 mm (a tolerance of ±3 mm would still be acceptable).

10. Refit the cutters on to the face of the block with a strip of paper under the back edge. Some operators do not

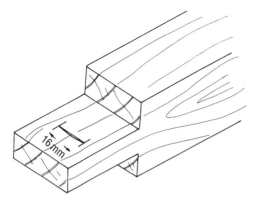

Fig. 4.56 Score mark left by spur cutter.

like to see paper behind a cutter, saying 'I will not rely on a piece of paper to save my fingers', but the fact is a strip of paper could do just that!

After a period of repeated tightening on to a block a cutter will start to bend very slightly. If by bending the cutter is allowed to come away from the front edge of the block, a gap will be created that will trap shavings and dust as the material is cut (**Fig. 4.57**).

As more waste is packed into the gap the cutter is forced to bend away even more, eventually resulting in

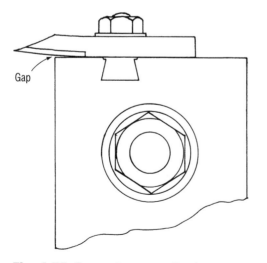

Fig. 4.57 Gap under cutter allowing waste to enter.

the cutter being bent back more than 90° in some cases (**Fig. 4.58**). Cutters being bent back more than 90° is not unknown, but more likely is splitting of the high speed steel tip from the mild steel, causing the ejection of metal parts.

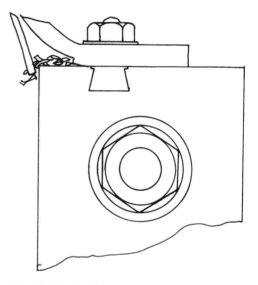

Fig. 4.58 Possible result to cutter.

The addition of a thin strip of paper under the back edge of a cutter is all that is needed to lean the cutter forward enough to prevent a gap from being present, thereby stopping the entry of waste material. The paper also reduces the chance of the cutter slipping out of position when tightening the securing bolts (**Fig. 4.59**).

11. Move the cutters.

- They should lightly graze or dress the face of the tenon. A rolling action of the block is used to ensure the cutters are level with the tenon face at their deepest point of cut and along their entire length.
- They should be approximately 0.5 mm away from the shoulder that was cut by the spur cutters.

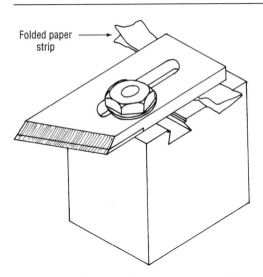

Folded paper strip

Fig. 4.59 *The use of paper behind a cutter.*

Having the tenon cutters too close to the shoulder will result in break-out or spelching along the shoulder. Too far away will leave too much material for the spurs to remove.

12. Tighten all securing nuts and/or bolts and make a final check of the setting of all cutters (tenon and spur).
13. Repeat steps 6–12 with bottom set.
14. Reposition all guards and safety devices and remove all tools and loose items from machine.
15. Pump start the machine and cut a test piece.

16. Check the finished cut for the faults shown in **Figs 4.60–4.63**.

This method of setting is commonly used but is only effective if the tenons cut before removing the cutters were parallel and without any of the faults above. If the setting is not particularly good before removing the old cutters it would be better to reset them on to a flat piece of freshly planed timber rather than a pre-cut tenon.

4.3.9 Machine adjustments

4.3.9.1 Tenon heads

The heads are mounted in slides for both horizontal and vertical adjustment, each with independent hand wheels for movement, and locks to prevent unwanted travel. A simple lead screw passing through a threaded bush provides the movement in both planes. An important feature on all machines is the stop that prevents the heads being wound together. It is usually a bolt set into the bottom head motor housing, and from time to time it may need adjusting (namely when new cutters are fitted).

As with all features on a machine the adjustments will vary depending on the make, type and size of machine. Machines with pillar castings will usually allow independent adjustment to each head in both vertical and horizontal planes as well as the function of linked

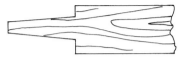

Fig. 4.60 Tenon cutters incorrectly set – angled instead of flat.

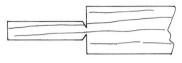

Fig. 4.61 Spur cutters set too deep.

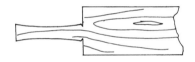

Fig. 4.62 Cutters ground with too much helical curve.

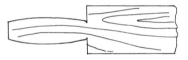

Fig. 4.63 Flat cutters fitted to shear cut block.

vertical adjustment (**Fig. 4.64**). A sprocket is simply engaged between the two lead screws to lock the heads together. This is used when the tenon thickness is correct (fitting nicely into the mortice hole) but its position is not correct (leaving a step on the joint between the two components).

Swan neck machines can also have linked adjustment, but this would eliminate the ability to pass long tenons between the heads and, as a consequence, they are very rare.

4.3.9.2 Scribing heads

Although scribing heads have independent adjustment (both vertical and horizontal), the slides they run on are in turn mounted on to the tenon head castings. Basically, if the tenon head is moved the scribing head goes with it. For this reason the tenon heads must be set before the scribing heads.

The purpose of a scribing head is to cut beyond the tenon shoulder to allow it to fit over a mould (**Fig. 4.65**).

Scribing is usually carried out by the use of a flush-mounted square cutter block with appropriately ground cutters bolted on to it. The cutters, just like those on the tenon heads, are made out of mild steel for the backing (or body) with small high speed steel (HSS) tips inserted at the end. It is the HSS that does all the cutting and they can blunt very quickly when working with dense,

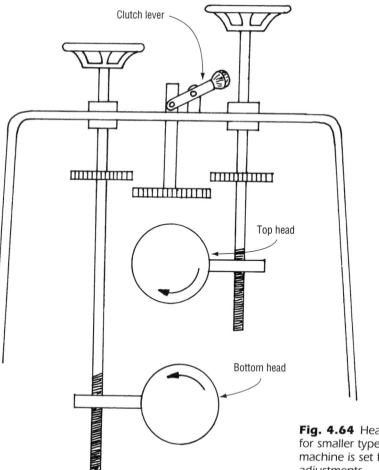

Fig. 4.64 *Head adjustment layout for smaller types of machine. The machine is set for individual head adjustments.*

abrasive materials. The block is also keyed on to the shaft. A small metal key is let into the shaft and the block is fitted over it to act as a driving device (**Fig. 4.66**).

An option now available and becoming ever more popular is to replace the square block for a serrated-knife circular moulding block, or even safety chip-

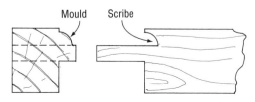

Fig. 4.65 *Scribe and mould.*

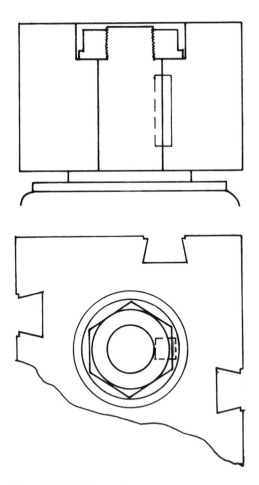

Fig. 4.66 Scribing block.

limiting blocks. A tooling company will recess the top, just like the square blocks, to allow the nut to sit below the surface. This will provide clearance for the tenon as it passes the head and prevent it from rubbing on the nut. Preground cutters including TCT (tungsten carbide tipped) are available for all types of block, provided in balanced pairs ready to run.

Hand wheels and locks are situated around the machine (again, depending on the type of machine) to aid adjustments .

4.3.10 Scribing cutters (setting)

After accurately balancing the cutters, nuts, bolts and washers that are to be used, setting can proceed. If a doubt is present relating to the balance or condition of any of the items, replace or rectify immediately (Chapter 5 explains about cutter balance).

To set a pair of cutters a setting template should be used. Ideally it should be marked out with projected 2 or 3 mm lines so that the intended cutter position can be plotted on to the template. (Chapter 5 gives more information about templates and cutter setting.) When a scribe has been produced with satisfactory results a template can be made where the cutters are scratched around to leave a line to set to in the future. One rule that must always be used when setting scribing cutters is to let the cutting point stand higher than the body of the cutter (**Fig. 4.67**). This will allow the

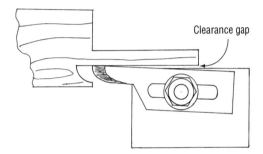

Fig. 4.67 *Setting a scribing cutter.*

cutter to reach beyond the shoulder to remove the timber but also provide clearance to prevent the cutter rubbing on the tenon.

4.3.11 Setting up to cut a tenon and scribe

The following procedure is a step by step guide to setting up for a standard tenon and scribe. It is, however, one of many ways which can be used, depending on personal preferences and the type of machine in question.

1. With the timber fully marked out, lay it on the machine bed and set the component cramp to hold it firmly in place.
2. Undo the locking nuts and adjust the heads so that they are in line with each other to cut square shoulders.
3. Slide the component towards the tenon heads until light contact is made with the spur cutters. The spur cutters should be touching the end grain of the component but still be able to revolve easily.
4. Adjust the height of the tenon heads so that, as the block revolves, the tenon cutters will be level with the marked lines on the end of the component.

There is no preferred order for setting the heads unless a permanently linked system is employed. This kind of system usually has two hand wheels for up and down adjustments. One will move only the bottom head when turned, whereas the other will move both the bottom and the top. It is therefore necessary to set the top head to the line first and then the bottom.

5. Select cutters, nuts, bolts, washers and a template for use on the bottom scribing head. The cutters, nuts, bolts and washers must be balanced.
6. Place the cutters on to the scribing block and butt the setting template up to the intersection of cutter and block.
7. Accurately position the cutters to sit on the lines of the template and tighten the nuts finger tight. Make sure that (a) the top of the cutter is higher than the block and (b) the cutter is slightly angled back. This will allow clearance, to prevent rubbing.
8. Tighten the nuts with a ring spanner or torque wrench and recheck the cutter positions against the template.
9. Advance the component into the machine so that a tenon can be cut. Start the tenon heads and cut the tenon.
10. Isolate the machine and feed the component past the tenon heads into the scribing area.
11. Raise the bottom scribing head to allow the tenon face to be very lightly touched by the point of the cutter. Too much contact will cause excessive wear on the cutting edge.
12. Adjust the in and out of the block to allow the scribing cutters to reach the marked out line beyond the tenon shoulder. This process can be awkward to achieve perfect results first time, but after a test cut adjustments can be made.
13. Prepare the machine for a test cut and feed the component through steadily.
14. Try the cut into the mortice hole and over the mould.
15. Make any adjustments until a correct fit of both tenon and scribe is achieved. Ideally the tenon should push into the mortice with moderate hand pressure, then it should hold its own weight when turned upside down. It should also separate by hand. The scribe should fit over a mould with no light showing through from the other side.
16. When satisfied with the joint, the

cut-off saw at the back can be set to trim the tenon to length. This will save valuable time later when the job is being glued up. Adjustments to the cut-off saw are by a hand wheel for in and out movements only.

4.3.2 Premoulded sections

Sometimes it is necessary to cut a scribed tenon on to a timber section that is already moulded. The chance of producing this tenon without break-out along the mould is almost impossible unless a premoulded backing piece is used. This is simply the reverse of the mould fixed to the fence with the timber pressed up to it. For maximum efficiency the backing must match the mould exactly. For a rebate this would involve pinning a piece of timber, equal to the sizes of the rebate, on to the plain backing piece. For a mould, however, the process is not quite so simple.

4.3.13 Making a premoulded backing piece

1. After fitting the cutters to the scribing head set the cutters as normal – tenon heads first, followed by the scribers. When satisfied with the setting run a test cut (ignore any break-out at this stage).
2. Make any adjustments to the cutters and heads as necessary until the scribed tenon fits into the mortice hole and pulls up snug to the mould.
3. Prepare a piece of timber to the same thickness as the mould and approximately 300 × 200 mm in length and width (**Fig. 4.68**).

4. Clamp the prepared timber into the machine lengthways so that the scribe can be cut on to the edge (running with the grain).
5. Using a sawing machine, cut off the scribed section (**Fig. 4.69**).
6. Pin the section on to the plain backing piece along with any other required pieces (**Fig. 4.70**).

The section to be tenoned must fit snugly into the backing in order to eliminate break-out totally. In some cases it is necessary to thickness the timber not to the size of the mould, as in **Fig. 4.68**, but slightly thicker than the section to be tenoned. This way the whole piece is supported by one moulded section, rather than having to piece together a backing piece (**Fig. 4.71**).

It may be difficult to get the components to pull up tight to the backing if it

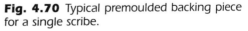
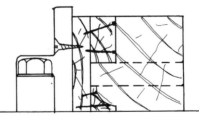

Fig. 4.69 Sawing off the scribed section.

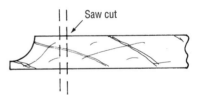

Fig. 4.70 Typical premoulded backing piece for a single scribe.

Fig. 4.71 Backing piece for double scribing.

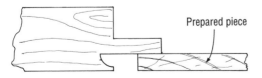

Fig. 4.68 Preparing backing.

is allowed to run the full length of the fence. It is normal practice to cut two 100 mm lengths of the backing and fit one at the front, where the cut takes place, and the other at the end on the component simply to keep it parallel.

4.3.14 Guards

As with all machines, the guarding of cutters and saws is extremely important and care must be taken to ensure that a machine's guards are set so as to prevent entry to any dangerous areas and moving parts.

Tenoning machines can be frightening because (a) the blocks are big, and therefore noisy, and (b) to allow the passage of material through the heads a large open area is present.

Although very few tenoners break regulations with regard to open cutters and insufficient guarding, there is room for improvement on these semi-automatically fed machines. One simple solution would be to fit two retractable covers to the sliding table that could be housed off the main casting. As the machine table is pushed forward to make a cut the cover will be unreeled to form a guard across the open blocks. As the table returns a lightly coiled spring would rewind the guard (working similar to a tape measure). This would ensure that no objects could be ejected from the machine towards the operator, but also and probably more importantly hands would not be able to wander into the cutting areas.

4.3.15 New developments

Modern tenoners are fast becoming popular due to the fact that they are so quick and easy to set up (**Fig. 4.72**). The

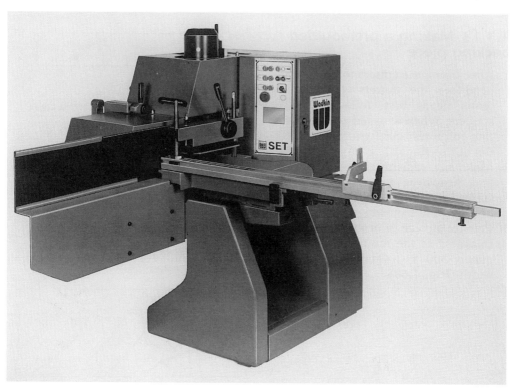

Fig. 4.72 Modern single-end tenoner.

scribing block is incorporated within the main tenon head. This itself is a circular cutting head with disposable TC tips for cutting the tenon and detachable cartridges holding HSS tipped cutters for scribing. The tenoning part of the block is very similar to a rebating block as used on a spindle moulder (Chapter 5). This is so not only in appearance but also in the fact that the cutting action of this machine is basically cutting a rebate on to the end grain. A useful addition to these machines is the dial-a-size counter display on all adjustment screws (**Fig. 4.73**). Once a job has been proved on the machine the readings of all the counters are recorded and, when the job needs

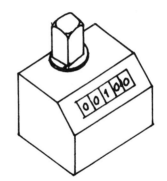

Fig. 4.73 Dial-a-size counter display.

setting up for again, the numbers are simply entered into the counters by turning the hand wheels.

Self-assessment

The following questions have been written around the previous text in this chapter. If you cannot answer any of the questions, simply restudy the respective areas. Good luck!

1. The two tenon heads of a swan neck casting machine are off set vertically to each other to :
   (A) allow the cutters to remove more waste
   (B) reduce the cutting impact
   (C) prevent the component jamming
   (D) prevent the blocks jamming.

2. A spring-loaded table stop is used for:
   (A) preventing the table from rolling into the blocks
   (B) preventing the table from moving while setting up
   (C) ensuring component length
   (D) applying pressure to hold the work piece.

3. A scribing block is recessed on the top to:
   (A) allow the tenon to pass it
   (B) make block removal easier
   (C) prevent the block from loosening when starting up
   (D) prevent resin from building up on the threads.

4. When fitting new tenon and spur cutters to a tenoning machine the spur cutters should be set:
   (A) 1mm behind the tenon cutters
   (B) level with the tenon cutters
   (C) on the bearing end of the block
   (D) in advance of the tenon cutters by 1 mm.

5. Helical curved cutters are popular on tenoning machines because:
   (A) they allow longer tenons to be produced
   (B) the shear action results in a smoother cut
   (C) the shear action creates smaller shavings
   (D) the cutters are less likely to chip.

6. There are two main types of machine casting on conventional single-end tenoners. Name them and give one advantage of each.

7. How can small pieces of timber be held safely and securely when cutting a tenon on to them?

8. Why is it sometimes necessary to make and fit a moulded backing piece before cutting tenons?

9. Why should a strip of paper be fitted behind a scribing cutter?

10. Linked adjustments can be made to most pillar-type casting machines. Explain how the heads can be linked together.

Crossword puzzle
Jointing machines

From the clues below fill in the crossword puzzle. The answers are all related to the chisel morticer or single-end tenoner.

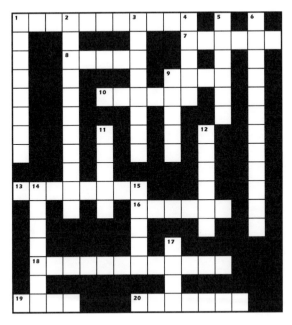

Across

1. Lansing or shoulder irons.
7. Smaller of the two types of tenoner.
8. Used to hold the work piece.
9. HSS ---- fitted to the end of a scribing cutter.
10. Small device for centring and holding an auger.
13. Name given to progressive cutting action.
16. Type of mortice that is produced when one end of the work piece is tilted up.
18. Hole where the waste is expelled from a mortice chisel.
19. See 18.
20. Shape of mortice chisel to allow a degree of clearance.

Down

1. Cut made beyond the shoulder to fit over a mould.
2. Shape of block with 50° cutting angle.
3. Cutters are bolted to this but not the bottom one.
4. Large blocks are made like this to reduce the length of shaving produced – also what happens to the wood when cut with no spur cutters.
5. A ring spanner is safer when changing blades due to the end being ------.
6. Moulded or plain – both prevent break-out.
9. Fits inside a mortice hole.
11. Work piece butts up to this on a morticer.
12. This cuts the mortice hole square.
14. Type of curve found on some tenon cutters.
15. Both a morticer and tenoner have one, but only one is sliding.
17. Dead, table and turn over.

5 Vertical spindle moulder

5.1 Vertical spindle moulder construction

5.1.1 Primary functions

The vertical spindle moulder is the most versatile of all woodworking machines and few workshops, if any, concerned with producing joinery or furniture are without one. Classed as its main functions are rebating, moulding and grooving, all of which can be done straight or curved.

5.1.2 Secondary functions

In addition to the primary functions, with slight alterations to guards, fences and tooling dovetailing, trenching, tenoning, comb jointing and, with the aid of jigs, small items including end grain work can also be carried out easily and efficiently.

5.1.3 Design and layout

From the manufacturer's and machinist's point of view the vertical spindle moulder is a very simple machine, comprising (**Fig. 5.1**):

- casting and bed;
- electric motor with pulley wheels; these can be with vee belt pulleys or flat belt pulleys, depending on machine make;
- spindle rise and fall hand wheel and locking lever, used to adjust the height of cutters for positioning tooling to suit marking out;
- spindle lock, used to hold the shaft while undoing the locking nut when changing cutting heads;
- brake, used to bring the machine to a halt after turning off. On older machines the brake will be operated by hand lever or foot pedal. Newer machines must comply with current regulations by having an automatic braking mechanism that will bring the cutting heads to a standstill within 10 seconds of being activated.

5.1.4 Block speeds

Most machines have two speeds, available through a stepped pulley driven by flat belts, but many modern machines incorporate a two-speed motor which, in conjunction with the pulleys, gives four speeds. This is usually designed so that it will only start in the slowest speed so as to prevent any damage should the wrong combination of block diameter and spindle speed be selected.

Cutter blocks with large diameters must be run at slow speed, whereas a small diameter cutter (such as a router cutter for trenching stairs) can be run at the maximum to give a reasonable

Fig. 5.1 Machine layout.

peripheral speed and thus efficient cutting.

Note that all tooling is stamped with its maximum running speed. Failure to keep within this limit will result in excessive vibration and could lead to cutters flying out.

5.1.5 Cutter blocks

a) Circular moulding blocks

This type of block (**Fig. 5.2**) is available in a wide variety of diameters and thicknesses. The most common one, referred to as the standard block, is 125 mm diameter × 50 mm thick. Most people refer to these blocks as whitehill blocks, although this is just a manufacturers' name. The cutters used on this type of block are approximately 4 mm thick and available in different lengths and widths to suit the job in question. They are made from solid high speed steel and can, therefore, be ground on all four edges. This greatly reduces the cost of tooling if the cutters

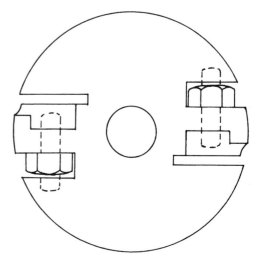

Fig. 5.2 Circular moulding cutter block.

are marked up and stored safely. In addition to the blanks available, where the machinist grinds the required shape, preground cutters are available whereby selection is made from a manufacturer's list. This type of cutter is aimed at the smaller firms with inadequate tool-room facilities or where an intricate mould is required to match up to preground scribing cutters that would take up valuable machining time if the machinist were to grind them.

The following safety factors need to be considered.

- Blocks must be kept in good condition. Any wear on the jaws or securing nuts and the block should not be used until it has been reconditioned.
- To help keep wear to the jaws down to a minimum the cutters must cover at least half of the jaw face.
- Cutter projection should not exceed approximately 15 mm (**Figs 5.3** and **5.4**).
- Cutters should always be used in pairs and both should be ground to the

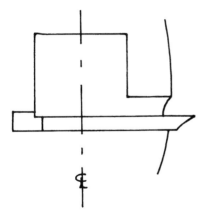

Fig. 5.3 Cutters beyond centre line.

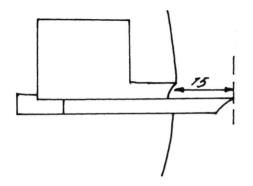

Fig. 5.4 Cutter projection not to exceed 15 mm.

same profile. Failure to do this would result in out of balance conditions.
- Never overtighten the securing nuts as this will stretch the threads.

i) Cutter balance

When a cutter block revolves at high speed, the block, cutters and components used to hold the cutters all come under the influence of centrifugal force. This is the action of an object to pull away from the centre as it spins. Centrifugal force is seen in many things in everyday life. A few examples are given below.

- As a motorcyclist enters a bend on the track he leans into the corner, if he remained upright the force would throw him off his bike.
- When a washing machine spin dries clothes the water is drawn off by centrifugal force. The clothes cannot actually go anywhere because of the drum wall but the water will pass through the holes in it. The faster the machine goes, the more force will be created and the more water will be removed.

For this reason both sides of a block must be identical in weight, balance, position and projection of cutters. This will involve the use of balancing scales to ensure equal weight and, once balanced, the cutters should be marked and kept in pairs. When mounting cutters it is important that they are set in the same position but on opposite faces of the block.

Figure 5.5 shows cutters set to opposite corners of the block. This will balance while stationary, known as static balance, but when revolving at high speed a twisting effect known as dynamic couple is present. This must be avoided at all times because, if it is allowed to continue, damage to the bearing is inevitable and a poor surface finish will result – the cutters here are equal **but not** opposite.

For a block to be dynamically balanced (balanced when running), the cutters must be set equal **and** opposite (**Fig. 5.6**).

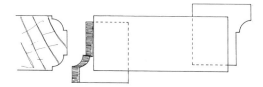

Fig. 5.5 Circular block with cutters set towards different edges giving static balance.

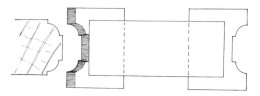

Fig. 5.6 Cutters set in dynamic balance.

With the cutters set like this, the force acting on one side of the block will be neutralized by the equal and opposite force on the other side, just as for balancing two items on a pair of weighing scales.

Cutters can be set accurately with the aid of setting or projection templates. A projection template is a formica (or similar) card with a series of lines scribed on to it, used to set the position of a cutter. **Figure 5.7** shows how the template must be made if it is to provide any useful accuracy. The divisions along line X are **true** measurements (usually 2 or 3 mm), whereas the divisions along line Y are **projected** measurements.

Projected measurements are required to allow for the fact that the cutter is off-set from the centre of the block. **Figure 5.8** shows that if the projection was ignored and the cutters were set away from the block by 29 mm, for example, then a rebate of 27 mm would be cut.

Procedure
1. Draw cutter block including centre line (C̸) and cutter face line from cutter seating (line Y).
2. Mark off 2 mm divisions onto line X[1].
3. Using a pair of compasses on the centre point of the block swing all the 2 mm divisions onto line Y.
4. Draw template outline and cut out to sit over the block (base of template to be on the first line).
5. Transfer all projected lines onto template.
6. Mark out actual 2 mm divisions along bottom line of template[1].
7. Square lines up onto template.

[1]*Note*: The actual 2 mm divisions are not drawn to scale.

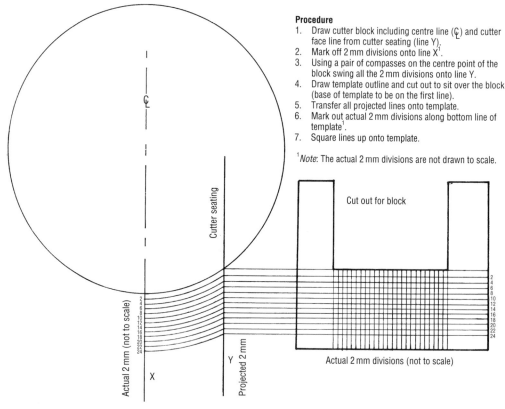

Fig. 5.7 Projection template.

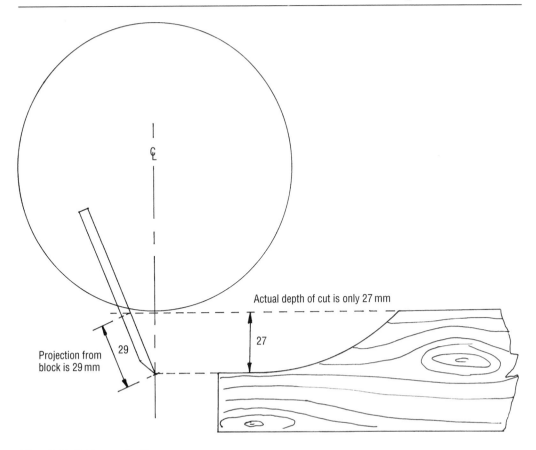

Fig. 5.8 Cutter projection.

Some machinists might guess the projection instead of using a template, but when cutting moulds the problem is far more difficult to guess. **Figure 5.9** shows how to develop the shape of a cutter to produce a mould.

It is also important to note that, if a mould cutter like the one in **Fig. 5.9** is to be used on a regular basis, requiring it to be removed and reset into the block, it should be set to the original position every time. If it is allowed to project from the block more than in the previous set-up, the mould will be cut slightly different. To see the difference, follow the procedure of developing a cutter with little projection and then over the top redraw it with an extra 20 mm projection.

To reduce the chance of supplying a customer with a differently shaped moulding (caused by different amounts of cutter projection), a setting template should be made (**Fig. 5.10**). This will allow the cutters to be removed and replaced time and time again yet still ensure that they are in the same place. Simply sit the template over the block, tighten up to the cutter face and scribe around the cutting edge with a sharp pointed tool (**Fig. 5.11**). It is best to do this at the end of a batch so that the cutting edge does not get damaged. Some companies have an indexed system of templates for all the jobs they do.

b) Chip-limiting safety blocks

With the introduction of new regulations, cutter blocks have been designed to meet safety standards by controlling the size of the chippings cut while

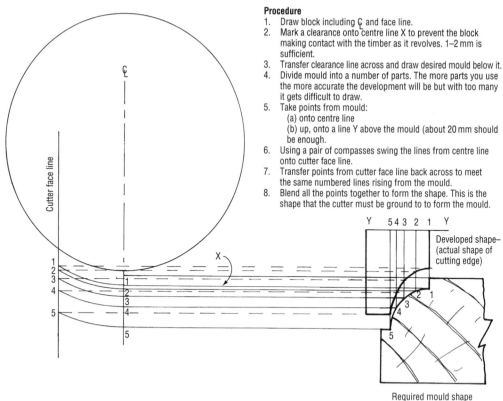

Procedure
1. Draw block including ₵ and face line.
2. Mark a clearance onto centre line X to prevent the block making contact with the timber as it revolves. 1–2 mm is sufficient.
3. Transfer clearance line across and draw desired mould below it.
4. Divide mould into a number of parts. The more parts you use the more accurate the development will be but with too many it gets difficult to draw.
5. Take points from mould:
    (a) onto centre line
    (b) up, onto a line Y above the mould (about 20 mm should be enough.
6. Using a pair of compasses swing the lines from centre line onto cutter face line.
7. Transfer points from cutter face line back across to meet the same numbered lines rising from the mould.
8. Blend all the points together to form the shape. This is the shape that the cutter must be ground to to form the mould.

Fig. 5.9 Cutter development.

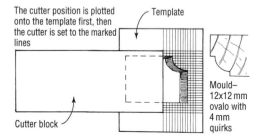

Fig. 5.10 Setting a cutter on a projection template.

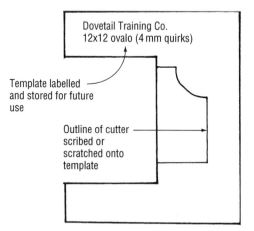

Fig. 5.11 Cutter outline scribed on to setting plate.

working (**Fig. 5.12**). The following two main safety factors are affected.

i) Kick back

This is when the material is thrown back towards the operator as a result of the cutters hitting into it. It can occur through uncontrolled feeding with insufficient pressure, contact with knots or

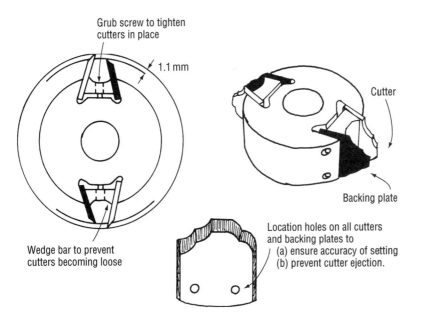

Grub screw to tighten cutters in place

1.1 mm

Cutter

Backing plate

Wedge bar to prevent cutters becoming loose

Location holes on all cutters and backing plates to
(a) ensure accuracy of setting
(b) prevent cutter ejection.

Fig. 5.12 Safety block and cutters.

foreign bodies, or when dropping on. The more a cutter projects from a cutter block the stronger the force will be as it digs deeper into the material.

By keeping projection to a minimum and providing the cutter with a backing or chip-limiting plate, the amount of force when kicked back is greatly reduced (**Fig. 5.13**). Basically, the timber is removed in thin slithers so that the depth of cut in any one revolution of the block will not exceed the limited chip size (approximately 1.1 mm).

ii) Hands making contact with cutters (Heaven forbid!!)

Owing to the way that we apply pressure when feeding into a machine (pushing towards the block), it is highly likely that in the event of kick back the operator will fall towards the revolving cutter block. By reducing the amount of cut in one revolution of the block far less damage will occur if contact is made.

A circular moulding block cutter with excessive projection will hook into flesh and instead of merely cutting a finger, it is possible for it to rip the finger off from the knuckle. With a chip-limiting block the flesh will be removed as thin, cleanly cut slices, meaning a far less serious accident (**Fig. 5.14**).

Note Relatively speaking, recorded accidents are few and far between and this paragraph is designed to show the advantages of new technology and not to scare the reader. If people are sensible

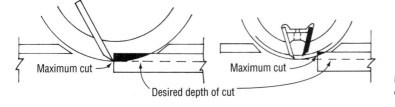

Maximum cut

Desired depth of cut

Maximum cut

Fig. 5.13 Maximum cut in each revolution.

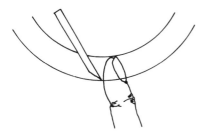

Finger enters into "Gullet area" and is ripped off at the knuckle!

Backing plate prevents finger going into gullet so only a thin slice is removed

Fig. 5.14 *How safety blades lessen the injury of an accident.*

and adhere to the regulations, accidents will not happen.

c) Solid profile blocks

The solid profile block is manufactured to a buyer's requirements and, because there are no loose pieces on it (such as bolts and washers etc.), there is no danger of cutters flying out (**Fig. 5.15**).

These are expensive to buy and, therefore, are usually used for mass production runs or a common section, e.g. ovalo mould. In addition to the initial outlay to buy the block, special grinding equipment is required to maintain the original profile.

d) Grooving blocks

In this category is a wide range of grooving heads and saws. Basically they are the same and do the same job, but different manufacturers fit different adjustments etc. Here are the three main types.

(i) Wobble Saw (Fig. 5.16)

Adjustment is made by moving the splayed collars to match up with numbers representing the desired width.

The obvious problem with this type is the fact that it produces a rounded bottom. However, manufacturers claim to have overcome this problem with new designs.

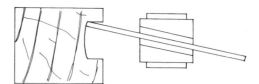

Fig. 5.16 *Wobble saw.*

(ii) Flat plate grooving saw (Fig. 5.17)

This type of blade is limited to cutting one size; therefore, it is available in a wide range of sizes.

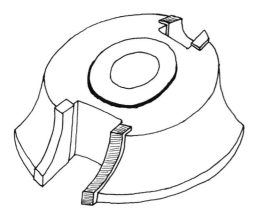

Fig. 5.15 *Typical solid profile block.*

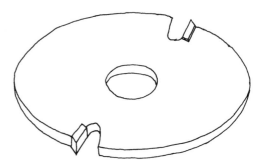

Fig. 5.17 *Grooving saw.*

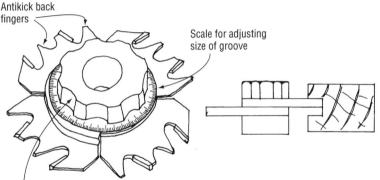

Antikick back fingers

Scale for adjusting size of groove

Comfortable grip to aid adjustments

Fig. 5.18 Segmental groover.

(iii) Adjustable, segmental grooving head (Fig. 5.18)

This is by far the most economical and efficient grooving head available. Adjustments in width are made by turning the calibrated screw and reading off the required setting. Each head is capable of an infinitely variable range of sizes between its maximum and minimum.

e) Rebating and variable angle bevelling blocks

These types of blocks are made specifically for rebating or bevelling, and with adjustment are capable of cutting an infinitely variable range of sizes (**Figs 5.19** and **5.20**). Owing to their flexibility they tend to be used very often and this would normally result in the cutters needing

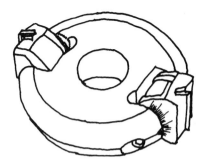

Fig. 5.20 Variable angle block.

regrinding regularly. To overcome this problem they are fitted with solid throwaway, tungsten carbide blades that are self-setting.

The shoulder or spur cutter is angled to give clearance and it is also set slightly in advance of the main cutter so as to sever the fibres and prevent break-out (**Fig. 5.21**).

The cutting block can be angled from 0° to 45° (in both directions) usually in

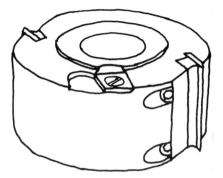

Fig. 5.19 Rebate block.

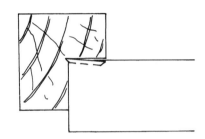

Fig. 5.21 Spur cutter on rebate block.

steps of 5° (**Fig. 5.22**). Clearance is built into the top of the cutter to allow rebates to be cut, although the finish is not as good on the underside without a spur cutter.

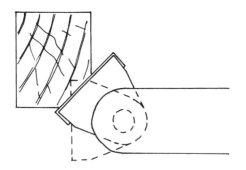

Fig. 5.22 Adjusting the angle.

5.2 Using a vertical spindle moulder

5.2.1 Setting up

Once the type of block and cutters is decided upon the machine can be set up.

As the shaft has the capability of being raised or lowered, to aid setting it is good practice to place a spacing collar below the block so that these adjustments are not restricted. This allows the shaft to be wound up without binding on the machine bed or reaching the end of its slide way (**Fig. 5.23**).

Note When using circular moulding (whitehill) blocks, it is possible to fit the cutters so that they stand below the block edge. For this reason it is good practice to fit a spacing collar between the block and the shaft base to prevent the cutters making contact with the bronze securing nut holding the spindle shaft in place.

It is also important to note that if the shaft needs adjusting in height, it must also be turned to prevent the belt coming off the pulley. If lowered or raised without turning the shaft by hand, the belt

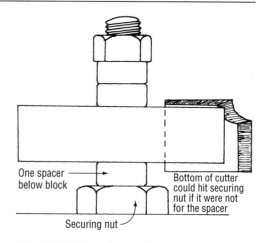

Fig. 5.23 Correct use of spacers.

comes off when started up as it jumps into place (**Fig. 5.24**).

In order to facilitate different block diameters, the bed has a series of steel rings that are inserted to close the gap around the cutter (**Fig. 5.25**). This is particularly useful when feeding short stock which could dip into the bed gap.

A gap is also present between the two fences so that the cutting part of a block can stick out. To close this gap a false fence known as a face board **must** be

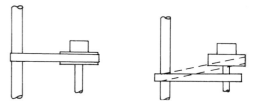

Fig. 5.24 Belt, before and after adjustments.

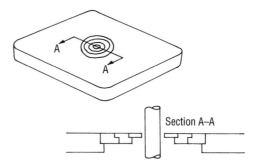

Fig. 5.25 Closing the gap in the bed.

used. It is usually made of waste timber 8–10 mm thick with a flat, planed surface on both sides and high enough to conceal the cutting area. Ten, 25 mm round wire nails are used to secure it on to the machine which should be inserted at different angles to each other (**Fig. 5.26**).

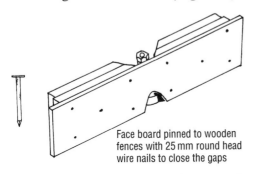

Face board pinned to wooden fences with 25 mm round head wire nails to close the gaps

Fig. 5.26 Closing the gap between the fences.

The procedure is as follows.

1. Fit the block on to shaft and tighten locking nut (left-hand thread to prevent it loosening when started).
2. Set the block height with either a rule measuring to a specific point or a setting card (**Fig. 5.27**).

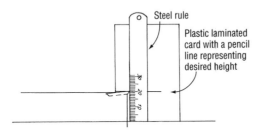

Steel rule

Plastic laminated card with a pencil line representing desired height

Fig. 5.27 Setting the cutter height.

3. Check the fences for being parallel and in line (see Section 5.2.1.2)
4. Nail the face board on to wooden fences with equal distance to each side. It is good practice to have the face board the full height of the fence, to prevent any open space when working.
5. Pump start the machine and slowly feed the fence into the revolving

block. This must not be jerked back as it will result in a hammering effect on the cutters, and most likely excessive tearing or break-out on the face board. Feed back until the cutter protrudes approximately 2 mm more than the required depth of cut by sighting up from above.

6. Turn off/isolate the machine and bring the fence forward by 2 mm to stop the cutters rubbing on the face board when started up. This will prevent blunting and reduce the noise level.

a) Reasons for using a face board

- To comply with regulations by closing in the cutter, thereby acting as a guard.
- To reduce the gap between fences to prevent timber falling in.
- To prevent dropping in any bent pieces.
- End grain work can be carried out.
- Clean cutting enables sharper edges and corners (chip-breaking effect).

b) Fence alignment

Before fixing the face board into place three checks must be made to the fences.

- Check for squareness to the machine bed. Shavings and chippings can build up behind and force them out of square (**Fig. 5.28**).

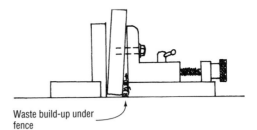

Waste build-up under fence

Fig. 5.28 Fence out of square.

- Check for parallel alignment. Again, build-up of waste material (behind the fences) can occur (**Fig. 5.29**).
- Check for level (**Figs 5.30 and 5.31**). Through repeated fine adjustments it is easy for the fences to be out of line with each other. The first two checks

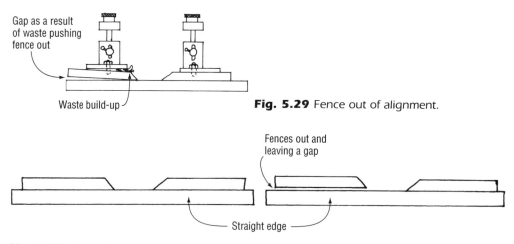

Gap as a result of waste pushing fence out

Waste build-up

Fig. 5.29 Fence out of alignment.

Fences out and leaving a gap

Straight edge

Fig. 5.30 Fences level.

Fig. 5.31 Fences out of level – work will drop in.

need only be carried out weekly, or if a problem is suspected. The last check should be made before pinning the face board on at every set-up.

c) Offset face board

When the work being done reduces the overall width of a component, the face board will need splitting and offsetting to give the component a bearing surface when being fed out, e.g. running a mould around the edges of a table top.

The dotted line in **Fig. 5.32** represents the original size of the top. If fed without any offset the back end would drop in after leaving the infeed fence (**Fig. 5.33**).

With the fences offset (similar to the bed arrangements of a surface planer, the component is supported throughout the cut with no possibility of it dropping in). This can also be used to give a different depth of cut, simply by adjusting the position of the infeed fence only (**Fig. 5.34**). The outfeed fence is set level with the part of the cutter with the least projection. This can be done by placing a steel rule against the fence to feel for light contact (**Figs 5.35–5.37**).

Note When feeding table tops (or any work that might be cutting end grain, such as the storm proofing of sashes), the following rules must be followed (**Fig. 5.38**).

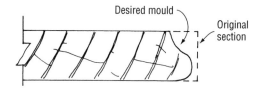

Desired mould

Original section

Fig. 5.32 Timber section with intended mould.

Gap for top to drop into

Fig. 5.33 No offset on outfeed fence.

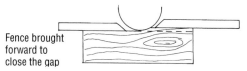

Fence brought forward to close the gap

Fig. 5.34 Fences offset.

1. Feed along the end grain on the first cut.
2. Turn the component anticlockwise to cut the other three sides.

By doing this any break-out on the end grain will be machined off in the following cut to leave a clean spelch-free finish.

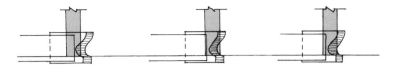

Fig. 5.35 Fence too far back – timber will drop in after leaving infeed fence.

Fig. 5.36 Fence too far forward – timber will get stuck as it reaches it.

Fig. 5.37 Correct – fence level with cutter.

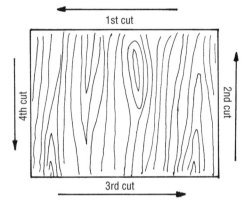

Fig. 5.38 Correct order for cutting around a board.

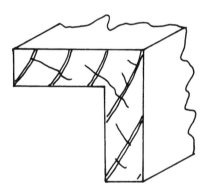

Fig. 5.39 Deep rebate.

d) Fence and bed pieces

The work of a spindle moulder is so varied that sometimes straight square fences need modifying to accommodate the timber. Here are three examples when such additions are required.

i) Machining deep rebates (Fig. 5.39)

In order to keep this piece stable while being fed through the machine, a saddle is pinned to the face board to fill the void and allow adequate pressure from the Shaw guards (**Fig. 5.40**).

ii) Cutting angled grooves or acute angles (Figs 5.41 and 5.42)

In both cases the cut must be made by tilting the stock.

This particular cut (**Fig. 5.41**) is made with a circular moulding cutter block where the cutter has been accurately ground to the required shape. Tilting the stock is made easy due to the bearing surface being the same angle as the cut, but a bevelled side pressure must be made (**Fig. 5.43**).

For this stock to be tilted a bed piece or saddle must be used (the amount of slope required will determine whether a

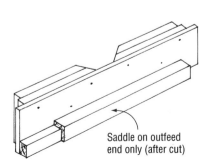

Saddle on outfeed end only (after cut)

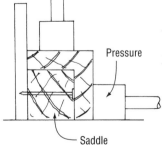

Pressure

Saddle

Fig. 5.40 Saddle filling rebate after being cut (conventional set-up prior to cut).

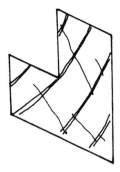

Fig. 5.41 Acute angled mould.

Fig. 5.42 Angled groove.

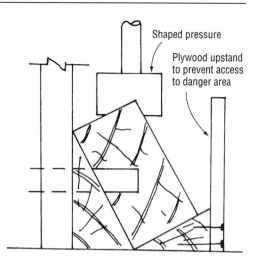

Shaped pressure

Plywood upstand to prevent access to danger area

Fig. 5.44 Set up to run angled groove.

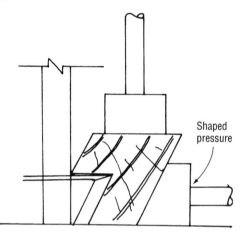

Shaped pressure

Fig. 5.43 Set up for mould.

complete saddle is needed or simply an angle block pinned to the fence (**Fig. 5.44**)). Unlike the deep rebate where the support is only needed after being cut, this type of aid must run the full length of the machine. Without this support the timber would be touching on two sharp edges and most probably would slide out of position. It is sufficient to use only a top pressure on this type of work as the piece cannot slide from the fence due to the supporting saddle.

Note In order to comply with current regulations, if no side pressure is used an up stand must be pinned on to conceal the cutters and prevent hands holding the timber wandering in the cutting area.

iii) Stopped work (dropping on)

When a door frame is being made which also has a glazed sidelight (combination frame), the rebate for the door will be on the opposite side to the light (**Fig. 5.45**).

In order to machine the head and cill in one setting-up of the tooling, the work needs to be fed in to cut one of the rebates without removing the mould further down. This is done by placing stops on the fence that prevent cutting too far. The sequence of operations is given below.

1. Set up the machine as normal and test-cut the rebate for size until correct.
2. Mark on to the fence where the cutter starts and finishes cutting by placing a steel rule on the fence and carefully turning the block until it is making light contact with the rule (**Fig. 5.46**).
3. Place the premarked-out timber on to the machine bed and position it so that the point that shows where the rebate should stop is next to the point labelled 'finish' and position a stop block at the end (**Fig. 5.47**).
4. With both top and side pressures in place, proceed to cut rebates on all suitable pieces (in this case the cills).
5. Remove stop and set it to the infeed end in the same way as stage 3.
6. For this cut, no side pressure can be

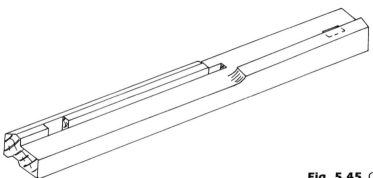

Fig. 5.45 Combination frame head.

fitted because the face that we will be feeding on to the revolving block is in the middle of the timber.

Note that, due to the speed at which the cutter revolves, there would be a definite possibility that the timber would be thrown out if it were not for the back stop (**Fig. 5.48**).

7. Place trailing corner on to stop and under top pressure.
8. With one hand at each end of the length of timber gently push the timber into the revolving block. Sometimes waste can build up between the fence and timber, keeping it from having the correct depth cut. If this appears to happen remove the timber and clear waste before reapplying.
9. Feed out with the aid of a push block or push stick.

In order to eliminate dropping on, some machines are fitted with reverse rotation. This allows one hand of the job to be fed normally and the other hand fed into the machine from the opposite end after the machine and tooling has been reset.

Another advantage of reverse rotation is that timber that would normally be fed against the grain can be fed with it to reduce any grain lifting. A good, sharp circular moulding block cutter would usually have a cutting angle of about 35°. This presents no problem when cutting straight grain but, if the grain is sloping or interlocked, it can run ahead of the cut and cause the fibres to split or tear

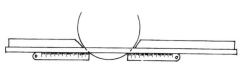

Fig. 5.46 Marking entry and exit of cutter.

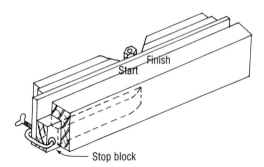

Fig. 5.47 Use of stop block (G cramped or nailed).

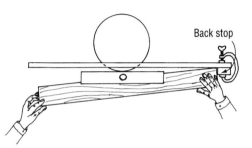

Fig. 5.48 Use of a back stop when dropping on.

out, thereby leaving a poor and broken surface finish (**Figs 5.49** and **5.50**).

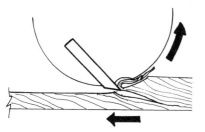

Fig. 5.49 Normal rotation – the fibres of the grain are plucked up to leave a broken finish.

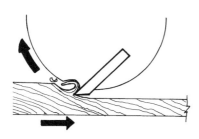

Fig. 5.50 Reverse rotation – the fibres are laid down, allowing the cutters to cut cleanly, leaving a good finish.

Note that when using reverse rotation the timber must be fed in the opposite direction to normal (towards the revolving block's direction of rotation).

Without reverse rotation the same problems with grain can be reduced by altering the sharpness or cutting angle (**Figs 5.51** and **5.52**). Where the grain is only slightly problematic a back bevel can be applied. This works by reducing the sharpness of the cutter, thereby pre-

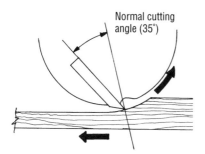

Fig. 5.51 Normal cutting angle.

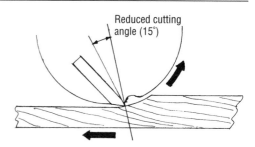

Fig. 5.52 Reduced cutting angle.

venting it from riving under the grain. With the feed speed slowed down as well, a good finish is possible. Where the grain is heavily sloping or interlocked a front bevel is required. It must be understood, however, that any tooling with a front bevel will require extreme caution due to the fact that kick back is far more likely. It is not advisable to feed by hand when back bevels are applied, but instead to use an automatic feeding unit (see Section 5.2.4). With a back bevel applied the cutting angle is reduced and the cutting principle takes on a scraping action (similar to a cabinet maker's scraper). The waste is removed as a much finer powdery dust rather than the usual large wooden flakes. Consequently the cutter cannot rive under the grain and split the fibres.

Usual practice is to slip-stone the bevels on to the cutters when required, although they cannot be taken off without a regrind.

Note that when a bevel is applied it makes the cutter blunt. This will in turn mean that more feeding pressure needs exerting, and that if the timber is left in contact with the revolving cutters severe burn marks will be present on the cut surface.

5.2.2 Safe operations

Wherever possible work the material over the cut. This will allow the material to act as a guard, thereby enclosing the cutter (**Figs 5.53** and **5.54**).

One exception to the rule is when cut-

Fig. 5.53 Correct set up allowing timber to guard the cutter.

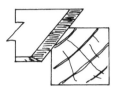

Fig. 5.54 Cutter over the top, exposing cutter.

Fig. 5.55 Cutting tongues.

ting tongues (**Fig. 5.55**). By working the cutter above, the tongue will be a constant thickness, even if the material varies in thickness. It is important that if this method of working is adopted sufficient safety devices are used to prevent the operator wandering into the exposed cutter.

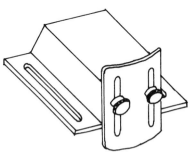

Fig. 5.56 Top guard with adjustable front plate.

5.2.3 Guards

When feeding any material on a spindle moulder the cutters must be enclosed and guarded as effectively as possible. Most machines will automatically guard the remote (rear) of the block by its fence arrangement or top guard (**Figs 5.56** and **5.57**).

Usually, the only exposed part is the working cutter (through the face board). This is guarded by the use of a Shaw guard (or tunnel guard; **Figs 5.58** and **5.59**).

Both top and side pressures must be sufficient to prevent material from moving away from fence or bed, but not so that feeding is difficult. To feed all the way through a Shaw guard, a push stick can be used. Avoid pulling the material from the outfeed end and never push with bar hands past the cutting head.

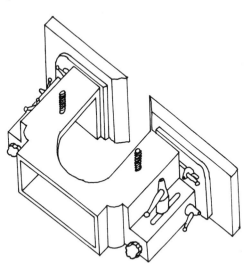

Fig. 5.57 Horseshoe fence.

5.2.4 Automatic feed units

However careful an operator is when feeding, there will always be times when the surface finish is affected by starts and stops or differences in feed speed. Continuous controlled feeding is only achieved by means of automatic attach-

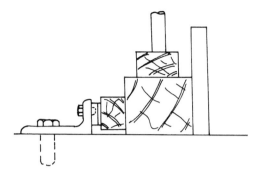

Fig. 5.58 Shaw guards set to provide a pressurized tunnel to feed through.

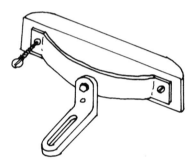

Fig. 5.59 Side pressure (Shaw guard).

ments (**Fig. 5.60**). These attachments can also increase production and cut down on the fatigue an operator may experience on production runs.

Each roller is independently spring loaded to allow for variations in material thickness. Although sometimes puzzling until familiar, adjustments can be made to position the rollers to suit almost any type of work and, with usually four different feed speed settings (eight including reverse), acceptable surface finish and production outputs are easy to achieve.

Some typical set-ups are shown in **Figs 5.61** and **5.62**.

Different opinions are voiced as to the correct position of the rollers in relation to the work. Usual practice is to keep the middle roller in line with the cutting head. This allows direct pressure while being cut, but also the work is held firm both before and after cutting (**Fig. 5.63**).

Rollers are also slightly inclined towards the outfeed fence in order to keep the material tight up to the fence.

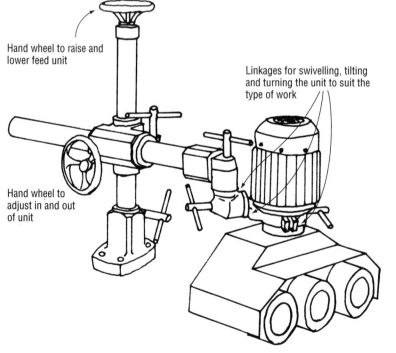

Hand wheel to raise and lower feed unit

Linkages for swivelling, tilting and turning the unit to suit the type of work

Hand wheel to adjust in and out of unit

Fig. 5.60
Automatic feed unit.

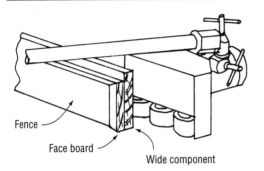

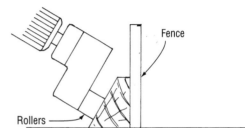

Fig. 5.62 Splayed or angled sections.

Fig. 5.61 Wide boards.

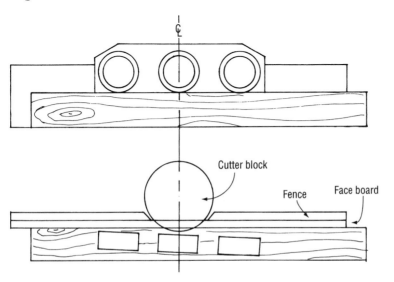

Rollers inclined to keep material up to fence

Fig. 5.63 Auto feed unit set with middle roller over cutting head and inclined towards the outfeed end of the fence.

(When feeding boards with the rollers pressing up against the fence, incline back rollers towards bed.)

Note Too much pressure will cause the back edge of material to lift on initial contact with the front roller. Too little pressure will allow material to wander while being cut.

Self-assessment

The following questions have been written around the previous text in this chapter. If you cannot answer any of the questions, simply restudy the respective areas. Good luck!

1. To ensure accuracy and consistency of moulds when running large batches, off which block would be the most suitable?
   (A) wobble saw
   (B) solid profile block
   (C) circular moulding block
   (D) variable angle block.

2. When cutting timbers with interlocked grain on a hand fed machine, what is advisable?
   (A) reduce the cutting angle
   (B) increase the cutting angle
   (C) use TCT cutters
   (D) slip-stone the cutters at regular intervals.

3. What are the cutters used in a circular moulding block for machining softwood made from?
   (A) high speed steel
   (B) mild steel
   (C) high carbon steel
   (D) stellite.

4. When straight fences are being used for work on a vertical spindle moulder, how must the gap between the fences be set?
   (A) to allow access for sharpening
   (B) to allow removal of chippings
   (C) reduced as far as possible by the use of a false fence or face board
   (D) increased as far as possible by removing the wooden fences.

5. When would the fences on a spindle moulder be set independently of each other ?
   (A) when cutting deep moulds and rebates
   (B) when using a flush top block
   (C) when reducing the complete width of a component
   (D) when the feed speed needs to be very slow.

6. Vertical spindle moulders with a two-speed motor for selecting different spindle speeds are designed to start only in the slower speed. Why is this so?

7. When adjusting the height of a spindle shaft what must be done to prevent the belt from riding off the pulley when using flat drive belts?

8. All tooling used on a vertical spindle moulder is stamped with a maximum running speed. Why is this so?

9. Safety blocks used on the vertical spindle moulder actually reduce kick back. Explain the term kick back and how the block reduces it.

10. When working a mould around a table top why is it so important to run the end grain first?

Crossword puzzle
Vertical spindle moulder

From the clues below fill in the crossword puzzle. The answers could be related to the vertical spindle moulder.

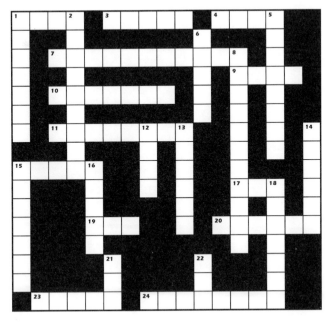

Across

1. (and 3) Used to hold the timber against the fence and bed.
3. See 1 across.
4. Best side and edge of timber.
7. Metric unit of measure.
9. Place where mould is put – face and ----.
10. A means of driving block but not flat.
11. Shape of whitehill block.
15. Decoration cut on to face corner.
17. Bend or distortion on end section; also used to drink out of.
19. Material used to make hard wearing cutting edges.
20. Plough cut with wobble saw.
23. Bonnet shaw or cage -----.
24. Even -------- is needed when feeding.

Down

1. (and 15) The machine's name.
2. Name commonly given to circular moulding cutter block.
5. Face board needed to mould this – it is usually cross-cut first.
6. Used to stop the spindle.
8. Formica strip used to set height of cutter.
12. Engaged to stop spindle turning.
13. Name of cut on the corner or timber – or money coming back from the tax man.
14. Shape of timber when width and thickness are the same size.
15. See 1 down.
16. How far the cutters go into the work piece.
18. The outline or shape of something.
21. The spindle shaft rises through it – the work rests on it.
22. Type of steel that circular moulding block cutters are made of.

6 Sanding machines

6.1 Introduction

Of all the machine operations covered in this book, sanding is probably the one that has seen the most rapid and dramatic developments. The demand for a better quality finish, higher output, lower tolerances of stock variation and the need to handle a wide range of materials and material sizes has seen machine companies produce some wonderful equipment. No longer does the term 'sanding' mean only the final task of smoothing a surface prior to accepting a finish coating. Abrasive planing and profile sanding are terms which are becoming more common in the industry.

This chapter provides a basic overview of sanding with the main, basic machines covered.

6.2 Wide belt sander

6.2.1 Functions

In recent years sanding machines have been subjected to extensive developments. The result is a massive variety of machines, each with specific use. The work of a wide belt sander (**Fig. 6.1**) is split into three main categories.

- *Precision or finish sanding*. This is the final sanding of a product following operations on cutting machines. Usu-

Fig. 6.1 Wide belt sander.

ally cutter marks, lumps, bumps and bruises are lightly sanded out to prepare the surface for surface coatings and treatments.
- *Abrasive planing*. Basically this is thicknessing with a sander. Poorly aligned joint surfaces and regularizing of stock is quickly and easily carried out.
- *Denibbing*. This is the process of

removing dust and raised grain from the surface of a lacquered or painted component in preparation for a second coat.

6.2.2 Design and layout

The three sanding operations listed above demand different working methods in order to be successful. Not only does the sanding belt need changing to give a finer or coarser cut, but also specific component parts of the machine will require different design and layout. The basic machine layout is shown in **Fig. 6.2**.

6.2.3 Contact roller

The roller is made of rubber, which will vary in hardness depending on the nature of work to be done on the machine. A hard rubber, for example, will have no give or flexibility to it and will therefore produce a very aggressive cut. This is perfect when the machine is being used to thickness or abrasive-plane a component. In fact, some machines are fitted with a steel roller so that no flex is

present. It means that the thickness of a component will be precisely and consistently the same from start to finish. Softer rubber could bounce or flex when making contact with knots and produce an irregular thickness. Soft rubber rollers were traditionally used for sanding veneered panels where a flex was essential to prevent sanding through the veneer, to expose the baseboard or glue pockets.

Whichever rubber is used, the roller is made with a series of spiralling grooves cut into the surface (**Fig. 6.3**).

The grooves provide several features:

- keeping the belt cool during cutting by carrying air between them;
- providing a progressive cut, which allows the belt to cut only when it is touching on the flats or lands of the roller; the life of the belt will be greatly increased as a result of this;
- helping to clean the dust off the belt, thereby reducing clogging;
- helping to grip the belt to produce a positive cut by preventing belt slip.

Different angles of the spiral grooves will provide different cutting properties. Steep angles will mean faster cutting.

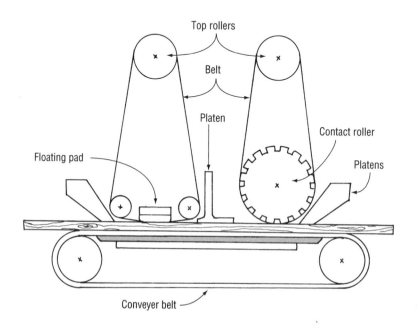

Fig. 6.2 Basic machine layout.

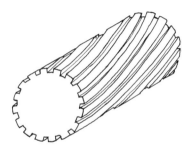

Fig. 6.3 Spiral grooved contact roller.

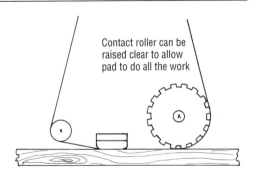

Contact roller can be raised clear to allow pad to do all the work

Fig. 6.4 Combination sanding unit.

Both the hardness of the roller and the angle of the grooves can be chosen to suit the type of work being done.

In preference to a soft rubber contact roller, when finish-sanding a floating sanding pad can be used. This is a hard-wearing block placed between two rollers and set low enough to make contact with the material. Again, the hardness of the block can be altered to suit the nature of work being done.

The main advantage is that the flat block produces a wider contact surface than that of a cylindrical roller. This will not cut as quickly, nor will it guarantee the accuracy of the component's thickness but, because it floats, it will produce a much better surface finish.

A machine made up of a combined unit, either as individual heads in one machine or as a combination head, would be capable of fine finish sanding or accurate thicknessing by altering the working height of the contact surfaces (**Fig. 6.4**).

It is worth noting that some operations carried out by sanding pads do present problems. A good example is when sanding doors where both with-the-grain and across-the-grain sanding takes place. Dust particles always build up between the belt and work piece and cause clogging. When sanding with the grain this does not show up, but when working across the grain the clogged dust leaves marks on the finished surface which stand out vividly. This just highlights the fact that one machine will be perfect only for a small number of operations. As pre-viously mentioned, a combination head or machine may be useful but you are then faced with the problem of which head should be first. The list of machine variations is endless, and for truly satis-factory results a graded sanding process is best where several processes take place in one pass. A three-headed machine, for example, could be arranged as follows.

- *First head.* A coarse belt is used on a hard rubber contact roller to bring the component to a standard thickness. Any joint irregularities, raised knots or glue spots are quickly abraded away.
- *Second head.* A medium belt is used to remove some of the roughness made by the first belt and to bring the thick-ness closer to the finished size.
- *Third head.* A fine belt running on a sanding pad is used to bring to the fin-ished thickness with a smooth and regular surface finish.

Machines can be ordered and built to suit a customer's every need with any number of heads, various hardnesses of contact roller or pad, bottom and top head arrangements to eliminate the need to turn the piece over etc.

6.2.4 Top roller

As the belt is moving around the contact roller it is kept taught by the height-adjustable top roller. Usually made of perfectly smooth metal, the top roller is pushed upwards to provide tension which allows the belt to cut efficiently. Compressed air is usually used to raise the roller, which is regulated to prevent over-tightening. In the event of a drop in air pressure the machine will automatically switch off to prevent any hazards. The top roller also acts as a tracking device, by swivelling to prevent the belt running off the rollers. All belts wander while running. This is known as oscillating, where the belt creeps from side to side. Without a tracking device the belt would wander in one direction until it came off the rollers. Tracking is a continual process of rocking first one way and then the other to keep the belt as close to the centre as possible.

Pneumatic tracking is where the belt is prevented from creeping off the rollers by the use of compressed air sensors. A jet of air is blown into a small receiver and, as the belt moves through the air jet, the supply into the receiver is cut off. From a central pivot the whole top roller will swivel to allow the belt to return.

Modern machines track the belt by the use of electronic eyes. These are beams of light shone from one point to a receiver opposite. As the belt breaks the beam of light a pneumatic piston is activated to swivel the top roller and make the belt return to its original position (**Fig. 6.5**).

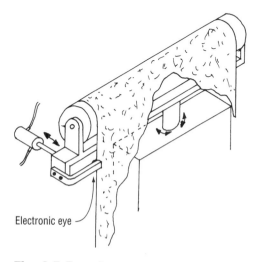

Fig. 6.5 Top roller arrangement.

Whatever tracking device is used, a common fault or problem is usually found at some stage during a belt's working life. When timbers are sanded the dust produced can build up on the abrasive grits that actually do the cutting. If the timber is resinous then more of this build-up will occur and produce small sections of belt which are clogged. These clogged areas can no longer efficiently sand the material and therefore rub the surface. As the belt rolls and oscillates from left to right a snake mark will be produced (**Fig. 6.6**).

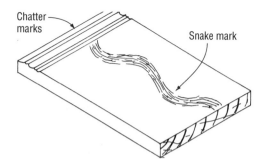

Fig. 6.6 Machine faults.

6.2.5 Platens (pressures)

A pressure bar known as a platen is positioned at each end of the machine to hold the material flat on the bed while being sanded. Both platens (front and rear) are adjustable to alter the amount of pressure they apply and both can also be locked in position with no movement at all. Although it very rarely needs checking, from time to time platen height must be checked in relation to contact roller height. If the platens are too low then excessive pressure will be exerted and feeding will not be consistent. Too little pressure will result in chatter marks as the work piece lifts off the bed during cutting.

6.2.6 Table

A rubber belt is tightly wrapped around two rollers. Tensioning and tracking devices are fitted onto the infeed-end roller, but once set they very rarely need adjusting. The surface is ribbed to provide grip and prevent the components slipping under load but periodically the surface may need roughening up. After many hours of use the belt will lose its tackiness and become glazed where it can no longer grip the work pieces. By gently and lightly raising the ribbed conveyor belt into a running sanding belt the glazed surface will be removed and the original conveyor conditions will be restored. This is also a useful technique with which to ensure that the table is running parallel to the contact roller.

Just like the platens, the table can be fixed in one position or spring loaded which will allow the table to 'float'. When abrasive planing, where the thickness of the finished piece is important, the distance between the contact roller and the table must always remain the same. This will mean that the stock can be brought to a regular thickness as it passes under the contact roller. Even if the component is tapered or has lumps in it to start with, the machine set with a fixed bed will remove the irregularities and produce a constant thickness of material coming out (**Fig. 6.7**).

Sometimes, however, the finished thickness is not the main concern. A veneered panel, for example, will have slight differences in thickness where a build-up of glue has occurred under the surface. To maintain a constant board thickness would mean that the veneer would be sanded through on the high spots to expose the original board below it, thereby making the component scrap. To avoid this the table is set to float (and the platens fixed). This will mean that as a lump approaches the first platen it will force the bed down to prevent the abrasive belt sanding through it. With the machine set in this way the same amount of material is removed from the top of the component (**Fig. 6.8**).

6.2.7 Feed speeds

The feed speed is obtained through an infinitely variable feed unit (see Chapter 3, Planing machines).

Wide belt sanders can be supplied with many optional extras such as:

- a brush roller – to clean off any surplus dust;
- a vacuum table – to hold the work piece flat while passing through the machine;

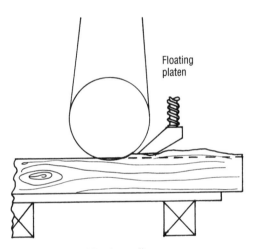

Fig. 6.7 Fixed bed sanding.

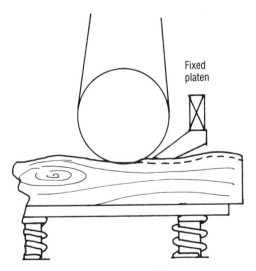

Fig. 6.8 Floating bed sanding.

- air jets – to clean the belts of dust and resin while running;
- segmental pressures – to accommodate more than one piece of timber being fed at the same time.

6.2.8 Sanding belts

Many factors are used to determine the suitability of a sanding belt for a particular job. A belt is basically a cloth or paper sheet with a layer of abrasive particles or grit glued onto it. Once a type of abrasive material has been selected, the size, hardness, toughness and quantity of particles can be decided to produce a belt suitable for a specific use.

6.2.9 Abrasive material

Three main types of abrasive are used for machine sanding.

- *Garnet.* This is a natural semi-precious stone which is reddish brown in colour. It is hard in its natural state and by heat treatment it becomes even harder. However, modern, high-speed machines soon knock the sharp points off the particles.
- *Aluminium oxide.* This is a synthetic abrasive produced from bauxite, a soft white rock which is the basic ingredient of aluminium. This is made into a molten state and mixed with coke and iron. As the air makes contact with the molten mixture an oxide skin forms which is crumbled to form the grit used as the abrasive. It is the ideal material for machine sanding as it is hard enough to withstand cutting impact, glue spots and abrasive timbers, thereby extending its working life.
- *Silicon carbide.* This synthetic abrasive is the hardest commercial abrasive. It is so hard that it is brittle, which means the long sharp points are easily broken off. Silicon carbide is best suited to hand sanding and will only

be found on machines if specifically requested for special jobs.

a) Grit size

As abrasives are crushed, the resulting fragments are graded ready for applying to the backing. To simplify, imagine that the grits are placed into a large bowl with holes in the bottom. All the small pieces will pass through the holes leaving just large particles. Below this first bowl are more bowls each with progressively smaller holes. Each bowl will collect one size of particle (**Fig. 6.9**).

Graded abrasives range from 12 (very coarse, large particles) to 1200 (extra fine, tiny particles). From this range wood-working industries tend to use 60–240 for standard stock sanding. For fine delicate work on veneered panels, or for denibbing a lacquered surface, up to 600 could be used.

b) Hardness

The hardness of an abrasive is whether the grits can actually cut into a material to sand or abrade it away. It is measured on a scale from 1 to 10. One is soft (talcum) and ten is the hardest (diamond).

c) Toughness

This is measured as a percentage and refers to an abrasive's ability to withstand wear as it rubs against a material.

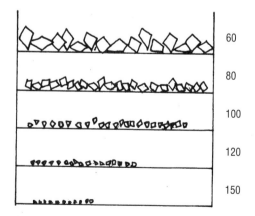

Fig. 6.9 *Grading the abrasive particles.*

d) Quantity

This is referred to as open or closed coatings. An open coat abrasive belt could have exactly the same abrasive grits on it as a closed coat, but the difference is that when a belt is classed as open only approximately 70% of the belt is covered with grit (**Figs 6.10** and **6.11**).

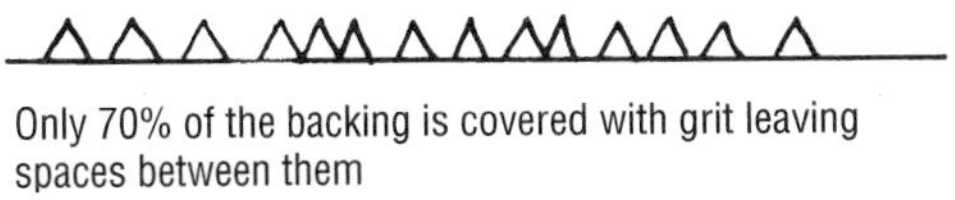

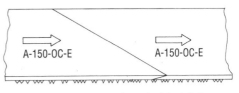

Only 70% of the backing is covered with grit leaving spaces between them

Fig. 6.10 Open coat abrasive.

Grit particles close together covering all the backing

Fig. 6.11 Closed coat abrasive.

Open coats are particularly useful when sanding resinous timbers, so that the stick dust can be collected in the open areas without clogging the belt (just like a large gullet on a circular rip saw blade). Closed coat belts are equally useful when rapid stock removal is required on dry materials such as particle boards.

6.2.10 Storage and handling

Belts are very fragile and failure to care for them properly will result in a very short run on the machine before breakages occur.

Ideally, belts should be opened out and hung near to the machine so that they can acclimatize to the atmospheric conditions around the machine. Belts stored near heaters will soon dehydrate and become dry. A dry belt has no flexibility to it and is therefore extremely brittle, thus requiring extreme care when handling and fitting to the machine. Conversely, a moist belt is very limp. Although it is less likely to break it can easily fold over or crease.

Once a belt has been fitted to a machine it should be run in for a minimum of 15 minutes before feeding any material through. This allows the belt to stretch fully and condition itself prior to working.

6.2.11 Belt joints

A good, strong joint is essential for successful working on machine sanders. Even then the joint is considered to be the weakest part of a belt and many breaks are as a result of the joint opening. Most belt nowadays tend to be joined by means of a lapped angle joint (**Fig. 6.12**).

A-150-OC-E   A-150-OC-E

Fig. 6.12 Lapped and angled belt joint.

The lap allows for a greater glue area to strengthen the joint and the angle creates a progressive cut to reduce cutting impact. Arrows are printed on the belt to indicate the intended direction to run. This will prevent the lap separating as contact is made with the component.

6.2.12 Backings

Two types of backing are commonly available.

- *Paper.* Grades C, D and E are usually available for machine sanding. The letters refer to the weight of the paper. Grade C is light weight (105–126 grams per metre square (g/m^2)) D is medium weight (126 to $158\,g/m^2$) E is heavy weight (over $218\,g/m^2$).
- *Cloth.* Cloth is usually graded X and J. X is classed as strong machine quality and J as medium.

Cloth is much stronger and more hard wearing, but for general sanding the abrasive wears out before the paper belt and is therefore of no benefit. Cloth is useful, however, when regular belt changes are carried out. It is the handling that

Fig. 6.13 *Linisher and disc sander.*

usually damages a belt, by creasing or tearing it.

A typical belt would be marked as

$$A - 150 - OC - E$$

where **A** is the type of grit (aluminium oxide), **150** is the grit size, **OC** stands for open coat and **E** is the backing paper weight.

6.3 Disc sander

A large metal disc coated with an abrasive sheet is driven by an electric motor and set into a machine casting which creates the working table or bed. Work pieces are laid down on the table and fed into the revolving disc which will quickly and easily abrade the unwanted material away. The bed is equipped with a fence to help in the locating and squaring of components and after adjustments both the bed and the fence can be independently set to create angles and compound angles (**Fig. 6.13**).

6.4 Linisher

This is basically the same as the wide belt sander in that it comprises a sanding belt that is tensioned and tracked around two rollers. The work pieces are laid down onto the moving belt to tidy or clean their surfaces ready for assembly.

Self-assessment

The following questions have been written around the previous text in this chapter. If you cannot answer any of the questions, simply restudy the respective areas. Good luck!

1. The contact roller of a wide belt sander may be covered with hard rubber. The hard rubber allows:
   (A) materials of varying thickness to be sanded
   (B) accurate thicknessing of the work piece
   (C) materials of varying width to be thicknessed
   (D) the same amount to be removed from the work piece.

2. Abrasive belts may be open coated. This means that:
   (A) the abrasive grits vary in size
   (B) the belt is punched with breathing holes
   (C) the grit covers only 70% of the belt
   (D) the belt comes in a long length ready to joint.

3. If the infeed pressure bar of a wide belt sander is fixed (not floating) this will:
   (A) bring sanded boards to an even thickness
   (B) enable frames to be sanded
   (C) hold down twisted boards
   (D) control the amount of material removed.

4. Which of the following would be most suitable for sanding chipboard panels to thickness?
   (A) open coated aluminium oxide
   (B) closed coated aluminium oxide
   (C) open coated silicon carbide
   (D) closed coated garnet.

5. A snake mark is something associated with:
   (A) ripple marks when feeding too fast
   (B) timber pulling away from the bed
   (C) the feeding conveyor not running true
   (D) abrasive papers with resin or dust build-ups.

Crossword puzzle
Sanding machines

From the clues below fill in the crossword puzzle. The answers are all associated with sanding machines.

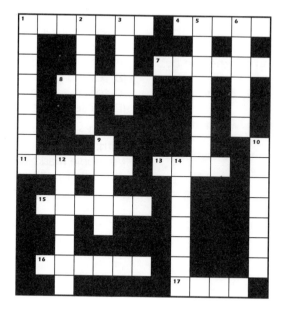

Across

1. Applied to strain the belt around the rollers.
4. The flats in between the grooves on a contact roller.
7. When the bed is fixed the pressures are ------ loaded.
8. 180, 120, 60 etc. used to describe grit.
11. Natural stone used as an abrasive.
13. 70% of belt surface is covered with abrasive grit.
15. Pressure bar.
16. See 2 down.
17. Particles of abrasive.

Down

1. Pneumatic or electric sensors to keep the belt from coming off the rollers.
2. (and 16 across) Cut into the contact roller to keep the belt cool.
3. Aluminium -----, result of molten mixture.
5. (and 14 down) Thicknessing on a sanding machine.
6. Removing dust particles from a lacquered surface.
9. Wrapped around the rollers and coated in abrasive particles.
10. Term given to a belt with a build-up of dust and resin.
12. One at each end of the conveyor belt.
14. See 5 down.

7 Waste extraction

7.1 Problem

In order to keep an area clean around a machine and also to comply with current regulations, extraction equipment is essential. Basically, air is sucked from the area close to where the waste is given off, and as the air is drawn off the waste is carried with it. Although hazards are present with waste in the form of chippings, it is the fine dust which we cannot always see that causes the real problems. Often it is so fine and light that before it can get into the suction path of the extraction unit it has dispersed into the surrounding atmosphere – the very air that we are breathing. The *Control Of Substances Hazardous to Health Regulations 1988* (COSHH) require that the air in work areas is clean. The air can be measured to see how much dust is present and to assess whether or not a health hazard exists (**Fig. 7.1**).

In very basic terms there are three sizes of dust. First, the big particles. These are breathed in but, thanks to the filters in our nose and throat, they cannot enter the lungs. They are usually cleared out at the end of the day by blowing your nose. Second, there are small particles. These manage to get through the filters in the nose and throat and make their way into the lungs. They are so small, however, that they can pass through the lungs and enter into the body's waste system for later ejection. It is the third type, the medium-sized particles, that cause the problem. These get into the lungs but are too big to escape and therefore build up inside the body. The problem is greatly increased when hardwoods are used for, two reasons.

- Hardwoods usually give off much more dust, probably because they contain less resin than softwoods and are therefore dryer.
- Hardwood dust is far more harmful than softwood dust. It is common to hear of allergies to hardwoods. Iroko, for example, is an extremely powerful irritant causing sneezing, restrictions to breathing and watering eyes within minutes of it being used. A hardwood named Obeche is well known in eye clinics for its corrosive nature. When dust particles settle in the eyes they cause severe swelling until the irritant is neutralized.

Another problem material is MDF (medium density fibreboard). Although it is made up of 90% softwood it is very dry and therefore produces vast amounts of dust while being cut.

The first line of defence against dust is efficient and well-placed extraction units. It may be worth investing in a particle mask to reduce the risk further, especially when hardwoods are being used. A point to emphasize the seriousness of the problem is this:

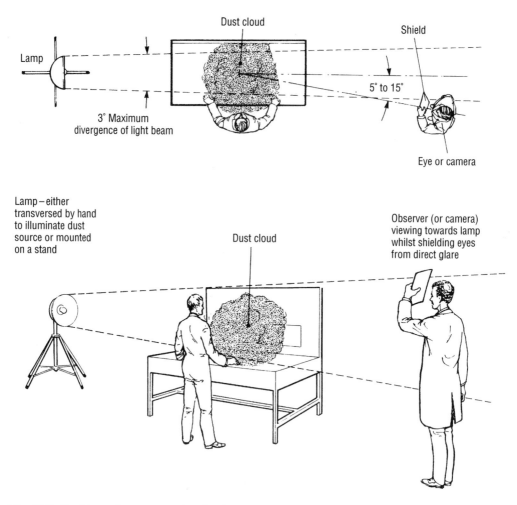

Dust cloud

Shield

Lamp

5° to 15°

3° Maximum
divergence of light beam

Eye or camera

Lamp – either
transversed by hand
to illuminate dust
source or mounted
on a stand

Dust cloud

Observer (or camera)
viewing towards lamp
whilst shielding eyes
from direct glare

Fig. 7.1 Dust lamp to measure dust in air.

A hardwood splinter in the hand can become infected within the hour, whereas softwood splinters very rarely get infected. To turn septic or infected is the body's reaction to the foreign object. If it does this when it is only a splinter in the hand, imagine what must be going on inside your body when particles get clogged in the lungs!

7.2 Types of extraction

7.2.1 Fabric filter modules

Most machines of this type work on the same principles, utilizing a powerful fan to pull air, loaded with waste (chippings and dust), into the extraction unit via an air inlet. As the air travels through the ducting or pipework and enters the unit settlement hopper the suction force will

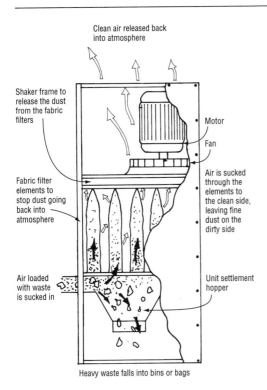

Clean air released back into atmosphere

Shaker frame to release the dust from the fabric filters

Motor

Fan

Fabric filter elements to stop dust going back into atmosphere

Air is sucked through the elements to the clean side, leaving fine dust on the dirty side

Air loaded with waste is sucked in

Unit settlement hopper

Heavy waste falls into bins or bags

Fig. 7.2 Fabric filter extraction unit.

drop and allow the big, heavy particles to fall from the flow of air into collection bins. The finer particles are carried up with the air into filtering elements which usually consist of fabric sheets. The air will progress through the fabric but leave the dust on the other side (known as the dirty air side). The clean, filtered air then leaves the machine and enters the atmosphere through an exhaust (**Fig. 7.2**).

Periodically the fabric sheets will need cleaning to remove the build-up of dust particles and allow air to pass through. This is done by shaking down. On older machines this requires the operator to move a handle on the side of the machine vigorously, but nowadays it is fully automated. As the machine is switched off an electric motor connected to it will start up and agitate the framework which car-

ries the fabric sheets. The dust will then fall away from the fabric into the collection bins.

7.2.2 Cyclone and baffle plate

A large, powerful fan sucks waste away from the machines towards the cyclone unit. The blades of the fan are designed to withstand pounding from large lumps of timber that make their way through the pipe. As the waste passes through the fan it is blown towards the cyclone at a powerful rate and if the pressure is not reduced the chipping will not be able to settle or drop into the collection points. To allow the force or pressure to drop, an air regulator known as a baffle plate is situated at the top of the cyclone and, by raising or lowering this plate, the air pressure can be adjusted. If the plate is lifted high above the cyclone then the air will quickly escape thereby making the waste fall instantly. This may also cause large amounts of dust to go with the air and pollute the atmosphere. The air flow must be set to allow the waste to spiral down the inside of the cyclone gradually and fall into collection bins. Any fine dust that still travels with the rising air can then be filtered before entering the atmosphere (**Fig. 7.3**).

Cyclones can also be incorporated with a built-in fabric filter unit to separate the chippings from the fine dust. This is particularly useful if the waste is to be either sold or burnt. The larger and heavier waste falls into the cyclone while the finer dust is carried in the air to the fabric filters (**Fig. 7.4**).

7.2.3 Sander dust

This must never be mixed with mill waste such as chippings because under

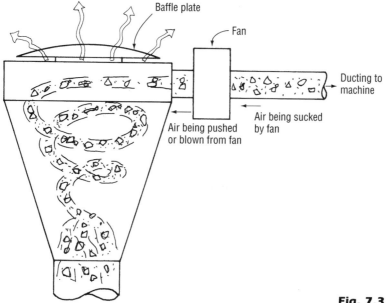

Baffle plate

Fan

Ducting to machine

Air being sucked by fan

Air being pushed or blown from fan

Fig. 7.3 Cyclone extractor.

Fine dust will carry on, past the cyclone to fabric filters before being collected into bins or bags

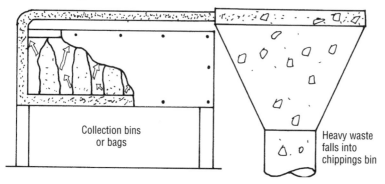

Collection bins or bags

Heavy waste falls into chippings bin

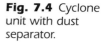

Fig. 7.4 Cyclone unit with dust separator.

certain conditions sander dust can be explosive. Every year factories are damaged or even destroyed by explosions. It is common practice to use the fabric filter modules for extracting sander dust with the only difference being the addition of a relief panel to reduce damage in the event of an explosion. The relief panel can be either an explosion door on the back of the unit (**Fig. 7.5**) or in ducting to the outside of the building. (**Fig. 7.6**). Whichever method is used, the area that the relief panel is in should be sealed off to prevent people wandering into the danger area.

Double-sided tape or magnets are used to secure door in position

Explosion Door KEEP CLEAR

Fig. 7.5 Explosion door.

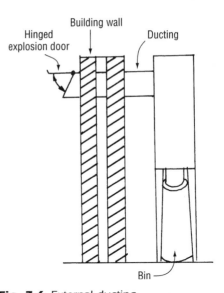

Hinged explosion door

Building wall

Ducting

Bin

Fig. 7.6 External ducting

8 Machine maintenance

8.1 Importance of maintenance and lubrication

Maintenance can be described as work undertaken to keep a machine in an acceptable working condition. This can be planned, where set lubrication charts indicate the frequency and location of the points to be lubricated, or unplanned, working with no schedule or manufacturers' recommendations. Basically, with unplanned maintenance, the operator will apply oil or grease to the respective areas as he sees fit.

In order that a machine can run efficiently and safely it is important to follow a systematic schedule, designed by the manufacturer of the machine. No amount of experience can equip you with the knowledge of how much grease a bearing can take, and even if you knew the amount it is doubtful whether it could be applied at the right time. Even a motor vehicle engineer will use a schedule when servicing a motor car. Remember – the manufacturer knows the machine better than you!

Failure to adopt a planned maintenance approach can result in:

- loss in production through unexpected breakdowns;
- high repair costs to replace damaged parts;
- shorter machine life due to either general failure or the cost of repairs and replacement being higher than a new machine.

The following are typical documents used for the servicing of machinery.

8.1.1 Card 1 – machine register

This gives the details of the machine to help identify it when carrying out a service. This is particularly useful when more than one machine of the same type is used. If spare parts are required or advice is sought about the machine these details are essential to ensure a speedy service (**Fig. 8.1**).

8.1.2 Card 2 – Lubrication schedule

This information is extracted from the machine manual (**Fig. 8.1**).

- *Code* refers to a series of letters placed on a simple line diagram of the machine. The letters represent the position of specific parts and therefore make it easier to locate them. For example, if 'A' refers to the machine bearings, there would be four 'A's on the diagram, each one representing a bearing.

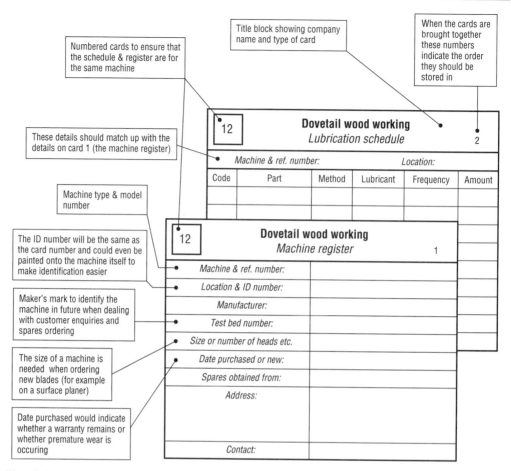

Fig. 8.1 Machine register card and lubrication schedule card.

- *Part* gives a name to correspond with the code.
- *Method* details how the lubricant will be applied.
- *Lubricant* gives details as to the type of lubricant to be used. Most factories have several oils and greases and the wrong choice could end up costly.
- *Frequency* explains how often the part needs charging with lubricant.
- *Amount* explains how much lubricant should be used. Avoid words like 'a few drops' or 'a small charge' or 'as required' because these will never be the same from one operator to the next. However, 'as required' is suitable for threads where the excess can run

off, but for bearings an actual amount must be stated.

8.1.3 Card 3 – servicing schedule timetable

This must be used in conjunction with the other two cards (**Fig. 8.2**). Whenever an element of servicing is carried out it must be recorded. This will prevent several people pumping grease into bearing within the 6 month period. As the greasing is carried out the sheet should be signed and dated with the details listed in the notes section. Notes are also used when a fault has been identified. By recording all the repairs that take place a

12	**Dovetail wood working**	
	Servicing schedule time-table	3

Machine & ref. number: SURFACE PLANER Location: M/C SHOP 1.

Date	Done by	Notes
31.7.97	SUPPLIER	Grease bearings in January. Full service prior to delivery
20.8.97	I. Bodgett	B,C,D,E carried out.
9.9.97	U. Scarper	B,C,D&E. Cutters changed-fault found on screw adjuster – new ones ordered

12	**Dovetail wood working**	
	Lubrication schedule	2

Machine & ref. no. SURFACE PLANER   Location: M/C SHOP 1.

Code	Part	Method	Lubricant	Frequency	Amount
A	BEARINGS	GREASE GUN	XTB40	6 mthly	4 Pumps
B	OIL NIPPLE ON BED SLIDES	OIL GUN	XT60	2 weekly	2 Pumps
C	Fence in & out ADJUSTER	OIL CAN	XT60	Weekly	2 SQUIRTS
D	BRIGHT PARTS	OILY RAG	Paraffin & oil	Weekly	wipe over
E	DRIVE BELTS	— CHECK FOR WEAR etc			

12	**Dovetail wood working**	
	Machine register	1

Machine & ref. number:	SURFACE PLANER ST400
Location & ID number:	M/C SHOP 1. – No. 12.
Manufacturer:	WADKIN
Test bed number:	ST4–0063–WP33Z
Size or number of heads etc.	400mm WIDE
Date purchased or new:	NEW: 31.7.97.
Spares obtained from:	NIKDUR M/c Co.
Address:	23,–29, Holbourne Road, Greenfield, High Sheaf. 02311 844255
Contact:	Salohain Nikdur.

Fig. 8.2 Completed machine register, lubrication schedule and servicing schedule timetable cards.

pattern may emerge linking the faults to some unexpected reason.

8.2 Lubrication

No matter how smooth a surface may seem there will always be a degree of roughness on close examination. Lubrication is the process of reducing the friction between two surfaces as they pass or travel over each other by placing a substance between them. **Figures 8.3** and **8.4** are exaggerated close-ups of metal surfaces. The roughness of surfaces acts like teeth interlocking and so prevent smooth travel. With a lubricant between the two surfaces travel is made much easier. The two main classes of lubrication are solid and fluid.

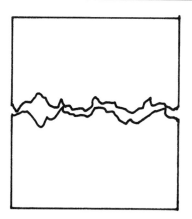

Fig. 8.3 Metal surfaces without lubrication.

Fig. 8.4 Metal surface with lubricant applied.

8.2.1 Solid lubrication

This is where a substance such as graphite, usually in the form of a powder or a suspension in oil, is placed between the surfaces. It will form a 'fish scale' layer due to the nature of graphite which will allow an easy slip. Solid lubrication is seldom used in our industry.

8.2.2 Fluid lubrication

This is where a film of fluid keeps the two surfaces from making direct contact. The only resistance then is the stickiness or viscosity of the fluid. Viscosity is a fluid's resistance to flow or its stickiness. It is created by the molecules that make up the substance passing over each other and creating friction. A fluid that is very thick, such as gear oil as used on heavy and slow moving parts, has a high viscosity. A thick oil will prevent either the oil being squashed out of the way on heavy parts or prevent it from dripping away from slow moving parts.

General-purpose oils and especially penetrating oils have low viscosities and are ideal for small intricate parts where access is limited.

Viscosity refers only to the thickness and has no bearing on the quality of a product. Some industries rely heavily on the viscosity and have made it compulsory to provide a numbering system, for instance SAE 30 as found on engine oil bottles. The number is achieved by measuring the time it takes for a cup of fluid to empty. This is all done under controlled conditions, but basically the longer the time taken for the cup to empty the more time will lapse and the higher the viscosity.

To simplify, relate it to water and treacle. Both are fluids but their viscosities are very different. Water, being thin, will flow easily and therefore has a low viscosity. Treacle, however, is very sticky and thick so it will flow much more slowly and have a high viscosity.

8.2.3 Types of fluid lubricant

A vast range of oils and greases are available, ranging from thin, watery oils to greases with a consistency of hard soap. Greases are in this category because they provide lubrication in the form of a fluid by bleeding oil into the bearing or softening under the heat. Greases are ideally suited to situations where it is essential to prevent the lubricant from dripping onto parts or products.

8.2.4 Methods of application

The most common methods for applying lubricants are via grease nipples. The five commonly found nipples are illustrated in **Fig. 8.5**.

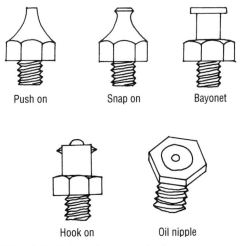

Push on   Snap on   Bayonet

Hook on   Oil nipple

Fig. 8.5 Types of grease nipples.

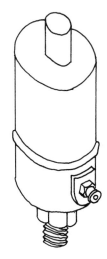

Fig. 8.6 Spring loaded grease cup.

All are simply a means of taking the lubricant from one place (the applicator) to another and all are non-return. This means that once the lubricant is pumped in it will not spill out when the gun is removed. The oil nipple uses a spring-loaded ball bearing in the end which moves clear to allow entry due to the pressure from the applicator. Once the applicator is taken away the ball will be pushed up to block the hole

8.2.5 Grease cups

The cup shown in **Fig. 8.6** is filled via the nipple at the base and the spring-loaded plunger then feeds a continuous flow of grease until empty. Any heat in the bearings will cause the grease to soften, thereby increasing the flow to reduce or correct the problem. This method is adapted from the screw-top grease cup which relies on a threaded cap being turned slowly, and as required, to force the grease into the bearing (**Fig. 8.7**).

8.2.6 Oil systems

Because oil is so versatile and easy to apply it is often wiped onto surfaces such as machine beds and fences or squirted from oil cans or even old washing-up-liquid bottles. There are, however, several

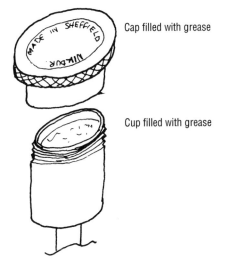

Cap filled with grease

Cup filled with grease

Fig. 8.7 Screw top grease cup.

specially designed systems for applying oil.

a) Oil cups

The cup is filled with oil after lifting the spring-loaded dust cap to allow access. The oil then slowly drips onto the respective parts (**Fig. 8.8**).

b) Direct oil feed pumps

These can be automatic or hand operated, but both work on similar principles. Oil is forced from the reservoir through

Fig. 8.8 Oil cup.

Both the previous systems are known as total loss systems. The oil is sent regardless of need and it is not recycled. Some systems, such as gear boxes, use the same oil over and over again. It is important that they are kept up to the required level to allow the parts to dip or splash into the oil and then lubricate the necessary parts. Also, the oil will only be used when the parts are moving or dipping into it and therefore the machine is running and is in need of lubrication.

8.2.7 Wick feeders

Some bobbin sanders make use of a wick feeder. This is a very simple principle where a supply of oil has a strip of cloth or cord draped into it. The oil then spreads along the material until a drip falls off the other end onto the moving part. Paraffin lamps work on this very principle by burning the paraffin as it spreads along the wick.

feed pipes to one or more distribution points. These tend to be found on the larger machines, such as double-ended tenoners, but can be used on thickness planers to oil the chain that drives the feed rollers (**Fig. 8.9**).

c) Gravity feed oiler

The flow adjuster can be altered to change the amount of oil allowed to pass through (**Fig. 8.10**). Usually a single drop per second is satisfactory. This can be assessed by watching the drops fall through the glass sight and turning the adjuster so that the needle is lifted up into the reservoir, thereby opening the gap and increase the flow.

If the machine has not been running for a while it might be worth increasing the flow for a few seconds to allow a flood to fill the distribution pipes and spread to the lubrication points. If the flow is left on while the machine is not in use the reservoir will simply empty onto the machine and surrounding areas.

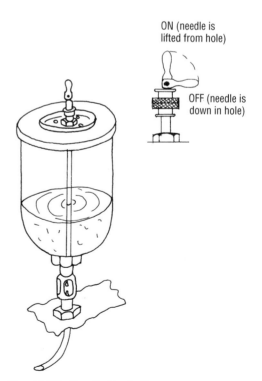

Fig. 8.10 Gravity feed oiler.

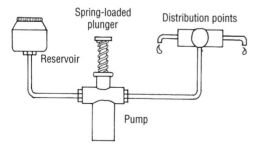

Fig. 8.9 Direct oil feed pump.

8.3 Servicing machinery

Routine maintenance and lubrication should be carried out with reference to the manufacturer's instructions, but also a thorough clean is required on a regular basis. A build-up of dust, chippings and waste is not good for the highly engineered parts of a machine. Even a simple screw thread will present problems by not winding into position if the teeth or threads are clogged up with waste.

The following list is a recommended approach for care and maintenance of a machine in addition to the manufacturer's servicing details.

- Ensure that the extraction system is working efficiently at all times. The more waste that is taken away by the extractor means less problems and work for you.
- Upon completion of a job and before setting up for the next one, brush any build-up of chippings to the extraction pipes.
- Once a week take time out to clean the machine and its surrounding areas. Check threads for build-up of dust and, if a build-up is present, use a wire brush to free the dust.
- Some parts may not be used very often during the everyday running of the machine. Check these parts to see that they still operate and have not ceased or rusted into position. Remember, even if you don't use the part today you may need it tomorrow and a blob of oil will keep it free and easy to operate.
- Refer to the schedule each week and check off the service requirements. If a bearing needs greasing every six months do not leave everything else for the same length of time. Be careful how you interpret the schedule as well. Often people will read off the schedule and merrily pump grease into a bearing every six months or as

prescribed, but if the machine has not been used for three of these months it could be considered that it is only three months, not six, since its last service. Conversely, the machine could be used in a factory working on shifts where the machine is working twice as hard and long as is normally expected. Some manufacturers write their schedules to be based on hours, i.e. 'grease every 300 hours'. This system is preferred but it is not always so easy to keep an accurate account of the working hours of a machine.

Too much grease is said to be worse than none at all because the grease cannot escape very easily from a bearing (unless the seals are opened). This is necessary to prevent the grease from dripping out in the event of the machine running hot and softening it. With an excess of grease it is packed in very tight and will solidify, causing the bearings to run hot. In severe cases it can cause the spindle to seize up or it has even been known for the ball bearings inside to split and break apart.

- Remove any build-up of resin from the beds, fences and other areas. A good scraper is useful here but it is also a good idea to soften the resin with a suitable component cleaner first. As in most of the chapters of this book, paraffin and oil will be suitable as the softening agent and a good scraper can be made out of an old mill file. Using a bench grinder with the rest set so as to grind the end of the file flat (not producing a point) lightly grind the end to a very slight curve (**Figs 8.11** and **8.12**). It is surprising just how efficient this tool will be at removing unwanted build-up on beds, fences and other component parts.
- Wipe all bright parts with an oily rag. Bright parts are bare metal parts which tend to rust or corrode if this thin smear of oil is not applied.
- Check drive belts for tension and wear. If one belt is worn or has broken

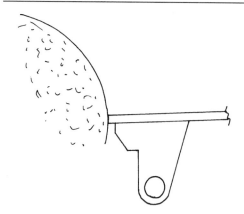

Fig. 8.11 Tool rest position.

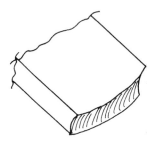

Fig. 8.12 Home-made scraper.

the whole set must be replaced. During the life of a belt it will stretch and get longer, thus creating the need for retensioning. If a single new belt were fitted to a pulley with two old belts only the new one would be driving the machine.

8.3.1 Sealed-for-life bearings

Modern machinery is often supplied with sealed-for-life bearings. This means that no direct maintenance will be required. During manufacture a high quality grease is packed into the bearings and these are then sealed to prevent either the grease getting out or any contamination getting in. This will cut down on maintenance time but, as with all machines, a wear check is still required periodically.

If bearings are worn excessively the machine will be very noisy under working conditions. This is because the bearings are not doing their job of holding the spindle shaft straight but are letting it vibrate. Tell-tale signs of worn bearings are either the noise, a build-up of heat or a wobbling, rocking movement when checked by hand. A spindle in this condition not only creates more noise, but also the surface finish of the component will be far worse due to the vibration.

Bearings are present on all machines and they are there to carry the load of and reduce the friction of moving parts. A spindle shaft with a heavy block on it would normally sag under the strain but the bearing, held in the bearing housing, keeps everything in position. Although there are many types of bearing they all work on the same principles. A set of hardened steel ball bearings or rollers are trapped between two sleeves each with a groove to accommodate the balls. The inner sleeve makes a tight fit over the spindle shaft while the outer sleeve does the same inside a bearing housing. As the shaft is turned the inner sleeve moves with it while the outer sleeves remains stationary. The ball bearings will now turn to reduce the friction between the two parts hence the name antifriction bearings.

The successful working of a bearing is only possible with the introduction of a lubricant to prevent the surfaces rubbing over each other. In most cases ball bearing grease does this job. Ball bearing grease is a synthetic or mineral oil that has been thickened and carefully manufactured to contain no impurities, such as acid or alkali, which could encourage corrosion and affect the performance of the bearing. Even when the bearings are not sealed for life, a cover plate is fitted to prevent contaminants entering.

A machine that is belt driven, such as a circular rip sawing machine, will have four bearings. Each bearing requires greasing so usually there will be four

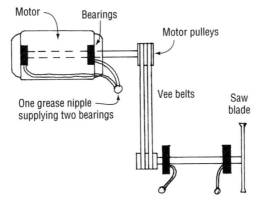

Fig. 8.13 Grease nipple and bearing arrangement

grease nipples but, if you cannot find four, check for the following. Assuming that the machine is not using sealed-for-life bearings which require no greasing, a single nipple could be feeding two bearing via distribution pipes (**Fig. 8.13**). This is particularly common where access is limited.

9 Calculations

9.1 Trade calculations

To work efficiently in any trade the use of numbers as a means of communication is essential. Whatever the problem, from working out the length or width of a component to the complex formula used to calculate the speed of a saw blade, numbers are moved around and worked on to produce an answer. All problems involve at least one of the four basic tools of mathematics.

9.1.1 Addition

This is the process of adding together a series of numbers to give a total.

Three receipts show the price paid for a series of items. You are to total them up in order to claim back the money as expenses.

1. Write down the prices keeping the decimal points in line with each other. This will keep units, tens, hundreds and thousands etc. in their correct columns:

hammer	£12.46
plane blade	£22.67
oil stone	£ 9.45
steel rule	£ 3.65

2. Add the right-hand column up and put the total below it: 6 + 7 + 5 + 5 = 23. If the total of any column is greater than ten, place the units (3 in this case) below the column and carry the tens (2 in this case) to the next column:

   ```
   12.46
   22.67
    9.45
    3.65
     ²
      3
   ```

3. Move to the next column and repeat step 2: 4 + 6 + 4 + 6 + 2 = 22. Put 2 down and carry 2 to the next column:

   ```
   12.46
   22.67
    9.45
    3.45
    ² ²
     2 3
   ```

4. Move to the next column and repeat step 2: 2 + 2 + 9 + 3 + 2 = 18. Put 8 down and carry 1:

   ```
   12.46
   22.67
    9.45
    3.45
   ¹ ² ²
   8. 2 3
   ```

5. Move to the last column and repeat step 2: 1 + 2 + 1 = 4:

   ```
   12.46
   22.67
    9.45
    3.45
   ¹ ² ²
   48. 2 3
   ```

Therefore, the total cost is £48.23.

9.1.2 Subtraction

This is the process of taking one number away from another to give a total.

If 33 pieces of timber are to be taken from a stack containing 292 pieces, how many pieces remain?

1. Write down the number you started with:

   292

2. Place below it the number you want to take away from it. Keep the decimal points in line and if no point is shown then assume it is at the end (e.g. 33.0 and 292.0):

   292
   <u>33</u>

3. Start in the right-hand column and take the number on the bottom from the one above it. To take 3 away from 2 would leave a minus or negative number so the sum must be altered:

   292
   <u>33</u>

4. 'Add 10 to the top and 1 to the bottom.' By using this simple rule the sum can be worked out without minus numbers. What this means is to make the 2 into 12 by putting a 1 in front of it:

   2 9$\overset{1}{2}$

   make the 3 into 4 by placing a 1 after it:

   3,3

5. 3 from 12 = 9

   2 9$\overset{1}{2}$
   <u>3,3</u>
   9

6. 4 from 9 = 5

   2 9$\overset{1}{2}$
   <u>3,3</u>
   59

7. 0 from 2 = 2

   2 9$\overset{1}{2}$
   <u>3,3</u>
   259

Therefore, the answer is 259.

9.1.3 Multiplication

Ten windows each need four hinges to hang the casements. How many hinges are needed in total?

This could be written as:

$$4+4+4+4+4+4+4+4+4+4=40$$

Simplified, we write this as 10 × 4. The × is saying 'of' (that is 10 of 4 = 40) written as a sum we put:

   10
   <u>4</u>

1. Multiply the 4 by 0:
   4 × 0 = 0 (4 of 0)

   10
   <u>4</u>
   0

2. Multiply 4 by 1:
   4 × 1 = 4 (4 of 1)

   10
   <u>4</u>
   40

Therefore, we need 40 hinges.

Multiplication becomes much more involved when bigger numbers are used.

If 120 houses each require 23 windows, how many windows are needed in total? This is written as (120 × 23).

1. Write it down as a sum:

   120
   <u>23</u>

2. Multiply the units together: 3 × 0 = 0

   120
   <u>23</u>
   0

3. Multiply the units by the tens: $3 \times 2 = 6$

  120
  <u> 23</u>
  60

4. Multiply the units by hundreds: $3 \times 1 = 3$

  120
  <u> 23</u>
  360

5. Now start in the tens column. To make the answer start under the tens we must place a 0 in the units column:

  120
  <u> 23</u>
  360
  0

6. Multiply tens by units: $2 \times 0 = 0$

  120
  <u> 23</u>
  360
  00

7. Multiply tens by tens: $2 \times 2 = 4$

  120
  <u> 23</u>
  360
  400

8. Multiply tens by hundreds: $2 \times 1 = 2$

  120
  <u> 23</u>
  360
  2400

9. Add the two rows together to get the answer: $(360 + 2400)$

  120
  <u> 23</u>
  360
  <u>**2400**</u>
  2760

Therefore, 2760 windows are required.

When multiplying numbers with decimal points, first write down the sum keeping the numbers in a line from the right-hand side. For example,

$$120.02 \times 21.4$$

would be written as

  12002
  <u> 214</u>

The decimal point is ignored until the sum is finished, then it is added by counting the number of digits to the right of it before we started. In this case 120.02 has two digits to the right of the point and 21.4 has one to the right. This gives a total of 3 places.

1. Multiply the units by the top row:

  12002
  <u> 214</u>
  48008

2. Place a 0 in the units column to make the answer start in the tens column and then multiply the tens by the top row:

  12002
  <u> 214</u>
  48008
  120020

3. Place two 0s in the columns from the right-hand side (one in the units and one in the tens) to make the answer start in the hundreds column:

  12002
  <u> 214</u>
  48008
  120020
  2400400

4. Add the three lines of the answer together:

  12002
  <u> 214</u>
  48008
  120020
  <u>**2400400**</u>
  2568428

5. Count the decimal point back in by three places from the right:

2568.428

9.1.4 Division

This is how many times one number will go into another. For example, if £900 bonus is to be shared out between 10 members of staff. We need to know how many times 10 will go into 900:

$$10 \div 900 = 90$$

therefore, each member would receive £90.

If 260 pieces of timber were to be divided into two piles, how many pieces would be in each pile? This is written as $260 \div 2$ (or 260/2).

1. Write the sum as shown:

$$2\,\overline{)260}$$

2. How many times does 2 go into 2: 2 into 2 = 1

$$\overset{1}{2\,\overline{)260}}$$

3. 2 into 6 = 3

$$\overset{13}{2\,\overline{)260}}$$

4. 2 into 0 = 0

$$\overset{130}{2\,\overline{)260}}$$

Therefore, 130 pieces would be in each pile.

This next example is more complicated. Divide 2960 by 32.

1. Write the sum into the box:

$$32\,\overline{)2960}$$

2. 32 into 2 will not go so carry the 2 over to the 9 to make 29:

$$\overset{0}{32\,\overline{)2960}}$$

3. 32 into 29 still will not go so carry 29 over to the 6 to make 296:

$$\overset{00}{32\,\overline{)2960}}$$

4. We need to know how many times 32 will go into 296:

$$1 \times 32 = 32$$
$$2 \times 32 = 64$$
$$3 \times 32 = 96$$
$$4 \times 32 = 128$$
$$5 \times 32 = 160$$
$$6 \times 32 = 192$$
$$7 \times 32 = 224$$
$$8 \times 32 = 256$$
$$9 \times 32 = 288$$
$$10 \times 32 = 320$$

9 times with 8 left over, so 32 into 296 = 9 remainder 8. Put the 9 on top of the box and carry the 8 over to the 0:

$$\overset{009}{32\,\overline{)2960}}$$

5. 32 into 80

$$1 \times 32 = 32$$
$$2 \times 32 = 64$$
$$3 \times 32 = 96$$

goes 2 times remainder 16:

$$\overset{0092}{32\,\overline{)2960}}$$

6. Carry the 16 to the next space. Anything added after the 2 is a decimal place so you can add a decimal point. Any empty space after the point is taken as a 0. By carrying 16 over we get 160:

$$\overset{0092_{16}}{32\,\overline{)2960.0}}$$

7. 32 into 160

$$1 \times 32 = 32$$
$$2 \times 32 = 64$$
$$3 \times 32 = 96$$
$$4 \times 32 = 128$$
$$5 \times 32 = 160$$

$$\frac{0092.5}{32\,\overline{\rule{0pt}{1.2ex})\,2960.0}}$$

Therefore, the answer is $2960 \div 32 = 92.5$.

9.1.5 Electronic calculators

The four basic calculations plus many other functions can quickly, easily and accurately be carried out on a calculator (**Fig. 9.1**). Some advanced models can even be programmed to calculate complex formulae. There are many hundreds of different calculators on the market and to use them properly it is advisable to read the manufacturer's instructions. Basic use is as follows:

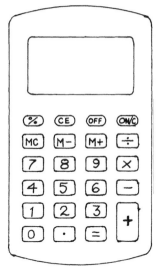

Fig. 9.1 Basic electronic calculator.

Function and sum	Buttons to press	Answer
Addition 23 + 19	[2][3][+][1][9][=]	42
Subtraction 23 − 19	[2][3][−][1][9][=]	4
Multiplication 12 × 9	[1][2][×][9][=]	108
Division 198 ÷ 16	[1][9][8][÷][1][6][=]	12.375
Percentage 62% of 420	[4][2][0][×][6][2][%]	260.4
420 plus 31%	[4][2][0][+][3][1][%]	550.2

9.2 Applying calculations to realistic situations

Often, in our industry, percentages will be used to calculate waste, savings, additions to existing orders etc. Percentage means per 100, so whatever we are doing (increasing or decreasing) we must base the sum on the original 100%. This could be written as

$$\frac{100}{100}$$

indicating that we have 100 parts out of 100. If we had to increase a quantity by 15% we would have the original amount which is 100% plus the extra 15%. This would be written as

$$\frac{115}{100}$$

Example: Increase 120 by 15%.
This is written as

$$120 \times \frac{115}{100}$$

by writing the question this way it looks much more complicated than it needs to be. To simplify the question divide the 115 by 100 before we start. Whenever you are dividing or multiplying by 100 (or 10, or 1000 etc.) simply move the decimal point by the number of 0s:

$10 \times 10 = 100$
decimal point moved once to the right

$10 \div 10 = 1.0$
decimal point moved once to the left.

Then

$115 \div 100 = 1.15$
$120 \times 1.15 = 138.05$

To decrease the same quantity by 15% we deduct the 15 from the original 100%:

$$\begin{array}{r} 100 \\ -15 \\ \hline 85 \end{array}$$

Then

$120 \times 0.85 = 102$

Example: A firm originally order 5500 metres of skirting board, but due to customer cancellations they have to reduce the order by 28%. How many metres are now required?

- $100 - 28 = 72$
- $5500 \times 0.72 = 3960\,m$

The company know that during the fitting and cutting of the skirting an allowance of 9% must be taken into account. What is the new total?

- $3960 \times 1.09 = 4316.4$

Therefore, 4316.4 m of skirting are required.

9.2.1 Bandsaw calculations

In practice the top wheel is wound up to its highest position then lowered by approximately 25 mm to allow for tensioning. A piece of string is then wrapped around the wheels and the length is marked onto the floor for future use. To calculate the length mathematically set the wheel position as before, then proceed as follows.

1. Measure between the wheel centres.
2. Measure the diameter of the wheel. The circumference is needed (the distance all the way around the outside of the wheel) and by measuring the diameter we can calculate it. Circumference = $\pi \times$ diameter, where π is the symbol used to denote 3.142 (or pi) which is simply the number of diameters needed to wrap around the circumference (**Fig. 9.2**).

For example, if a circle measures 100 mm across the diameter, the circumference would be $\pi \times$ diameter or $3.142 \times 100 = 314.2$ mm.

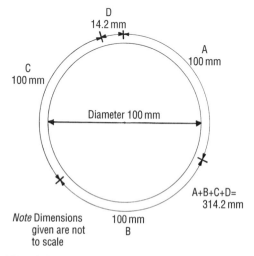

Note Dimensions given are not to scale

Fig. 9.2 Diameter and circumference.

The formula for bandsaw length is

$$L = (\pi \times D) + (2 \times C)$$

where

    L = length
    D = diameter
    C = distance between wheel centres.

Example: A bandsaw has 760 mm diameter wheels which are 1100 mm apart at their centres after allowing for tensioning. Calculate the length of the blade.

$$\begin{aligned}
L &= (\pi \times D) + (2 \times C) \\
&= (3.142 \times 760) + (2 \times 1100) \\
&= 2387.92 + 2200 \\
&= 4587.92 \text{ mm}
\end{aligned}$$

This can be converted to metres by moving the decimal point 3 places to the left.

Therefore, the length of blade required is 4.587 m.

9.2.2 Speed of cutting tools

The cutting speed of wood working tools is extremely important with particular reference to the following.

- *Cutting efficiency* – too slow will not allow the wood to be cleanly cut.
- *Safety* – too fast could cause cutters to fly out from the block.
- *Surface finish* – too slow would leave large cutter marks on the cut surface.
- *Power consumption* – too slow means an increase in power needed to force the cutters through the material.
- *Life of the cutting edge* – too fast will cause the cutter to rub on the material which will cause overheating and a blunting effect.

Cutting speeds are affected by many factors, such as the belts slipping, thereby not generating the intended speed, the feed speed of the material being too fast, or simply taking off too much cut.

Cutting speed refers simply to the speed at which the outside edge of the tool spins at and is measured in metres

per second. The following are recommended speeds and it must be realized that very few, if any, machines will hit the speed exactly. If even the motor creates a small amount of lag or loss in efficiency the cutting speed will be altered.

circular saw blade	50 mps
cutter blocks	25–45 mps
bandsaw blades	35 mps
abrasive belts	30 mps

The formula for cutting speed is

$$PS = \frac{\pi \times D \times RPS}{1000}$$

where

    PS = peripheral speed (the cutting speed)
    D = diameter
    RPS = revolutions per second the block.

The 1000 is simply there to convert mm to m for the final answer.

Example: A surface planer has a cutter block measuring 95 mm in diameter and revolves at 100 rps. Calculate the peripheral speed.

$$\begin{aligned}
PS &= \frac{\pi \times D \times RPS}{1000} \\
&= \frac{3.142 \times 95 \times 100}{1000} \\
&= \frac{29849}{1000} \\
&= 29.849 \text{ mps}
\end{aligned}$$

Sometimes it may not be the cutting speed that we need to find. If a saw is supposed to run at 50 mps we would usually know this. Instead, it could be that we need to work out the diameter of blade required for a particular machine. To work this calculation the formula for peripheral speed will need altering to find the missing part. This is known a 'transposing the formula'.

Example: A circular sawing machine with a motor running at 35 rps is required to produce a cutting speed of 50 mps. What diameter of saw blade is needed?

The first step is to transpose the formula.

Write out the formula for peripheral speed:

$$PS = \frac{\pi \times D \times RPS}{1000}$$

Put D to the front of the equation to represent the unknown value:

$$D =$$

Because a value (the D) has crossed to the other side of the equal sign (=) the remaining equation must be inverted (turned upside down):

$$\frac{D}{PS} = \frac{1000}{\pi \times RPS}$$

Finally place the PS in the original position of the value it has changed places with:

$$D = \frac{1000 \times PS}{\pi \times RPS}$$

Now the numerical values can be inserted:

$$D = \frac{1000 \times 50}{3.142 \times 35}$$

$$= \frac{50000}{109.97}$$

$$= 454.66945$$

This would be rounded up to the nearest commercially available blade size.

Therefore, a blade of 450 mm in diameter would be required.

9.2.3 Planing of timber

Surface finish, whatever the product, is very important especially on planing machines. Some people argue that it does not really matter as the product will be sanded later anyway. However, a good finish now will save time and money later.

When a piece of timber is fed through a planing machine the waste is removed chip by chip as the cutters revolve and the timber is fed forward. This will always produce what is known as a rotary finish where a series of waves or cutter marks will be visible on the cut surface. The size of these cutter marks will be determined by the speed that the material is fed (feed speed), the speed at which the cutter block revolves (revolutions per second) and the number of cutters in the block that actually leave a mark on the surface. It must be noted that on a surface planer block with two, three or four cutters only one cutter will be registering on the finished surface. All four will be removing chips of wood but, unless you are extremely lucky, one will be set slightly in advance of the others, even if it is only by a hundredth of a millimetre. For this reason it must be assumed that only one cutter per block is registering on the cut surface. The one occasion where more than one cutter will be registering is when the cutters have been jointed. This is a process of bringing all the cutters into one circle by lowering a fine stone, held in a special device, onto the cutters as they revolve. Jointing is not allowed on hand-fed machines due to the increased risk in kick back and very few thickness planers are fitted with the jointing attachment.

The formula for working out the cutter mark pitch is

$$P = \frac{F \times 1000}{RPS \times 60 \times C}$$

where

P	=	pitch
F	=	feed speed
RPS	=	revolutions per second
C	=	number of cutters registering a mark on the cut surface.

1000 is to convert metres, as used in the feed speed, to millimetres.
60 is used to convert seconds to minutes in the RPS.

Example: A thickness planer has been fitted with a four-knife block which has not been jointed. If the block revolves at 60 rps and the feed speed is set at 12 m/min what will the pitch of the cutter marks be?

$$P = \frac{F \times 1000}{\text{RPS} \times 60 \times C}$$

$$= \frac{12 \times 1000}{60 \times 60 \times 1}$$

$$= \frac{12000}{3600}$$

$$= 3.33$$

Therefore, the pitch of the cutter marks will be 3.33 mm.

Perhaps of more importance to the machinist is the ability to calculate the feed speed. This way the product can be made with a suitable pitch mark for the end use. Suitable pitch marks are as follows:

furniture and cabinet making 0.5–1.5 mm
high class joinery (polished)   1–2 mm
joinery   1.5–3.5 mm
carcassing and sizing work   3.5–6 mm

Example: Calculate the required feed speed of a thickness planer to produce a pitch of 1.5 mm using a four-knife, unjointed block revolving at 125 rps.

First, transpose the formula:

Write out the standard formula for pitch:

$$P = \frac{F \times 1000}{\text{RPS} \times 60 \times C}$$

Put *F* to the front of the equation to represent the unknown value:

$$F =$$

Because a value (*F*) has crossed to the other side of the equals sign the remaining equation must be inverted (turned upside down):

$$\frac{F}{P} = \frac{\text{RPS} \times 60 \times C}{1000}$$

Finally place the *P* in the original position of the value it has changed places with:

$$F = \frac{P \times \text{RPS} \times 60 \times C}{1000}$$

Second, insert the numbers from the question and perform the calculation:

$$F = \frac{1.5 \times 125 \times 60 \times 1}{1000}$$

$$= \frac{11250}{1000}$$

$$= 11.25 \, \text{m/min}$$

This would be rounded to the nearest feed speed on the machine without exceeding it.

If the block had been jointed then all four cutters would be registering and a much faster feed speed would be allowed:

$$F = \frac{P \times \text{RPS} \times 60 \times C}{1000}$$

$$= \frac{1.5 \times 125 \times 60 \times 4}{1000}$$

$$= \frac{45000}{1000}$$

$$= 45 \, \text{m/min}$$

Answers to self-assessment multiple choice questions

Chapter 1
1 = B, 2 = C, 3 = A, 4 = C, 5 = C

Chapter 2
1 = C, 2 = D, 3 = B, 4 = C, 5 = D

Chapter 3
1 = D, 2 = C, 3 = B, 4 = D, 5 = B,
6 = B

Chapter 4
1 = B, 2 = C, 3 = A, 4 = D, 5 = B

Chapter 5
1 = B, 2 = A, 3 = A, 4 = C, 5 = C

Chapter 6
1 = B, 2 = C, 3 = D, 4 = B, 5 = D

Index